AuthorHouse™
1663 Liberty Drive
Bloomington, IN 47403
www.authorhouse.com
Phone: 1-800-839-8640

First published by AuthorHouse 10/16/2009

ISBN: 978-1-4490-3050-6 (sc)

Printed in the United States of America
Bloomington, Indiana

This book is printed on acid-free paper.

UNITED STATE OF AMERICA

2009

EDUCATION IS POWER

OBAMA

STIMULUS EDUCATION PLAN (OSEP)

THE VOICE OF THE AUTHOR
Msc.-ing Feuzeu Yomsi
Raymond Jacquino

About Author

Msc.-ing Feuzeu Yomsi Raymond Jacquino, born in Cameroon in1967. Grew up and attended high school in Yaounde, Cameroon.

In 1991, My late father " **Yomsi Dieudonne** and my late mother " **Mekoue Rose**" provided a scholarship that enabled me to leave Cameroon to pursue further studies in **URSS**. After studying the Russian language for six months, I attended the Moscow state aviation University and obtained a **Master Degree** with honors in June 1997. My field of specialty was "automated systems of information processing and controls" in the department of Automated systems controls of spacecraft complexes. My theses was titled:
Sub system of planning of the Automated control system .

I served as a tutor in computer science with the collaboration of the Cameroonian Association (CamSam) in Moscow in 1997 . A year after I leave Moscow for Germany to continue my studies in Germany which began with the learning of the German language
From 1999 to 2004 , I also served as a tutor in computer science at the Christian University of Kiel (Germany) with the collaboration of the Cameroonian Association of Kiel (KSV); then at the Hamburg Harburg University (Germany) with the collaboration with the Cameroonian Association in Hamburg (ACH) and lastly a PhD research Student at the Hamburg Harburg University in Germany. As assistant I recorded the cabling of high frequency control of the injector area of the **TESLA TEST FACILITY** at **DESY** research center in Hamburg (Germany).
In 2004 I interrupted the P**hD r**esearch program to join my family in **USA.** One achievement of my life is to have been the co-founder of the Cameroonian Association of Greater Philadelphia(ACPE) and the president since 2006 to 2010 . I ' am grateful for the opportunity to serve the community: we created the Philadelphia -Douala Sister city Committee. During this period, I also obtained a life insurance certificate in Philadelphia.

As volunteer for the **Obama** campaign and inspired by his election as president, it is my pleasure to express words my Unqualified satisfaction , I feel to write this book , to share my ideas ,experiences and vision. By the Grace of God, I am devoted to contributing to the change by putting in place a network for tutoring system across America, Europe, Africa ,,,,,,,, to help students and challenge the Education system in the world

I would like to used this opportunity to thank my wife **Martine Feuzeu** for her support and love ,my **two sweet Daugthers Kelly Feuzeu** and **Velina Feuzeu** , my sister in law **Foppa Ivette** and her husband **Daniel Foppa** , **Ashley Foppa** and **Cyrian Foppa . Personal thanks to** my friend **Martin Bangha** for his intellectual contribution. Thanks to my family in Cameroon , to ACPE members, to **Guy Moyo , Nyoumsi Guy**.

To God be the glory .

Msc.ing Feuzeu Yomsi
Raymond Jacquino
7015 Hazel Ave
Upper Darby,Pa
19082, USA
email:feuzeu58@hotmail.com
tel:610-734-12-67,484-612-76-12

Barack OBAMA
PO. Box 802798
Chicago ,IL 60680

December 02,2008
-----0----

Proposal for a stimulus Education Plan

STIMULUS EDUCATION PLAN (SEP)

Introduction

I have designed here a stimulus education package that can " jolt the education into shape. This could be named OBAMA STIMULUS EDUCATION PLAN (OSEP) if eventually adopted. The OSEP will create millions of jobs, challenge our education system, motivate our students of all categories across the country .OSEP will help thousands of students to become successful in all academic areas by putting in place a network of tutoring to assist students.

Education is the key of success. Without it you may only get so far but in the end you are limited. Dropping out or failing to complete limits a person's lifetime chances of progress and achievement. The victory alone is not the change we seek .It is the only way for me to make a change and I believe this stimulus education plan will help people back to work ,open doors of opportunity for our kids , restore prosperity and to the American dream .

Accept my stimulus Education plan.

Thanks

Sincerely,

Msc.ing Feuzeu Yomsi
Raymond Jacquino

I. AUTOMATIZATION OF THE OBAMA STIMULUS EDUCATION PLAN

A.FUNCTIONAL STRUCTURE OF THE SYSTEM OSEP

B.RELATIONAL DIAGRAM

Msc.ing Feuzeu Yomsi
Raymond Jacquino

I.AUTOMATIZATION OF THE OBAMA STIMULUS EDUCATION PLAN (OSEP)

a. Introduction

**The key features of this automatization system include two majors systems. sub-systems: sub-system planing and operational sub-system.
we will focus on the sub-system planing : that means the OPTIMIZATION OF THE OBAMA EDUCATION PLAN.**

b. OPTIMIZATION OF THE OBAMA STIMULUS EDUCATION PLAN

The key features of the program include:

Motivation: OSEP will motivate all students across the country to become successful in all academic areas of the curriculum.

Assistance: Because of the high demand of tutoring in the environment OSEP will create a network of assistance to students

Minimization: Our goal is to minimize the academy difficulties face by students in school

Maximization: Eventually maximizing the success , increasing the level of knowledge, etc ...

c. The relation between a student and the tutor via the administration of the OBAMA stimulus education plan

VI. A.STUDENT CONSULTATION AND APPRECIATIONOF THE COMMITTEE

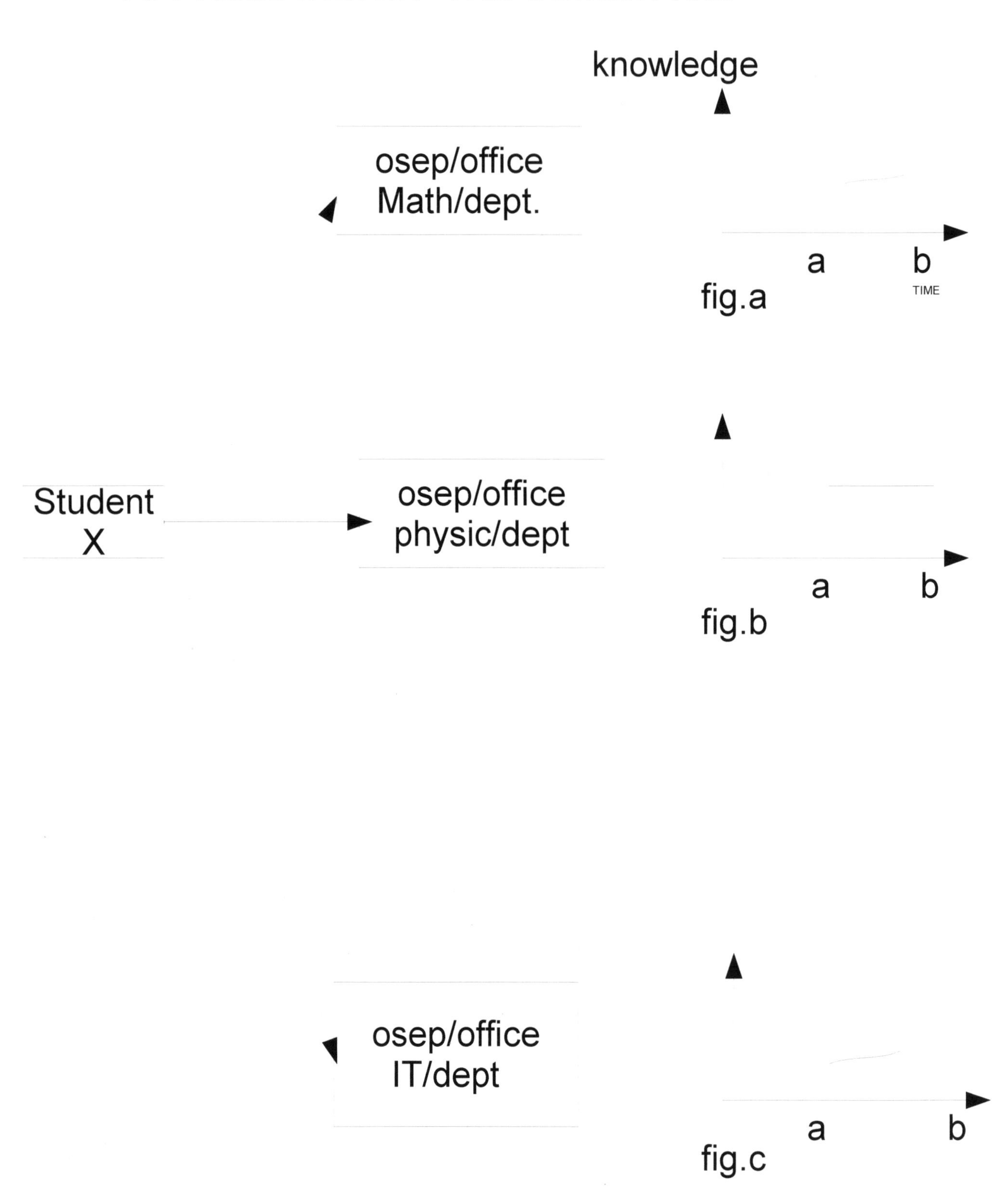

Msing. Feuzeu Yomsi
Raymond Jacquino

VI . STUDENT CONSULTATION AND APPRECIATION OF THE COMMITTEE

a. Graphical Interpretation

b. Description

Student x	**request assistance in mathematic and obtained**	**fig a. Good**
Student x	**request assistance in physic and obtained**	**fig b. Constant**
Student x	**request assistance in IT.**	**fig c. excellent**

VI I. PARAMETERS OF THE SITUATIONAL REPORT OF EVERY OFFICE

IDOSEP:Identification number of the OBAMA STIMULUS EDUCATION PLAN office

STAOSEP:State where located

CTY: Name of the city

NOSEP: Number of the students who receive a consultation at OSEP office a day,a week, year

SOSEP: Summary of students satisfy with this OSEP at that office

SnOSEP:Summary of students not satisfy with this OSEP

Ntmath:Number of tutors in mathematic committee of the OSEP in that office

Ntphy:Number of tutors in physic committee of the OSEP in that office

NtIT:Number of tutors in Information Technology committee of the OSEP in that office

etc....

MsucOSEP:The average of the students who became successful with the OBAMA STIMULUS EDUCATION PLAN in that office

MnsucOSEP:The average of the students not satisfy or unsuccessful with the OSEP

Ti: Total number of hours of consultation offered to student , etc ....

V. THE OSEP TUTORS SELECTION AND FUNCTIONAL DESCRIPTION

A. RECRUITMENT OF CVTUTORS

The OSEP will make announcement via Internet , TV,Radio,Newspapers.

The OSEP application will be on line or at the OSEP office available with all conditions request and the OSEP selection will follow up to see who is better qualify for the job

B. FUNCTIONAL DESCRIPTION OF THE OSEP

1. **Tutors of the mathematic committee of the OBAMA STIMULUS EDUCATION PLAN will help students member of the OSEP become successful in mathematic area by offering solutions their academic difficulties.**

2. **Tutors of the physics committee of the OBAMA STIMULUS EDUCATION PLAN will help student members solutions their academic difficulties in physic**

..

n. Tutors of the Informations Technology committee will help students become successful by minimize in Information Technology difficulties face by them .

3. The service will be at one of the OSEP office or on line

II. INFORMATION PROCESSING STEPS

step0 Student enumerate all academic difficulties he/she faces and contacts the OBAMA STIMULUS EDUCATION PLAN office. The student will also be encouraged to indicate the efforts and attempts made to resolve the enumerated problems.

step1 The OBAMA STIMULUS EDUCATION PLAN office immediately Identifies and contact the appropriate tutor committee to set up an appointment

step2 The OBAMA STIMULUS EDUCATION PLAN tutors provide the appropriate support to the student in need .

II .PARAMETER DESIGNATION

The OSEP will constitute or assemble a network of tutors sub divided into various key disciplines(mathematics, physics,... information technology, etc.) that is mathematic tutors committee , a physic tutors committee, information technology tutors committee etc...

IV STUDENTS NEEDS AND RESPONSIBILITIES

1. Registered with OSEP , to be OSEP member
2. Study to become successful
3. Identify or enumerate he/her difficulties
4. Give evidence of personal attempts to resolve problem
5. Be ready eventually to become an OSEP tutor to help others in need
6. etc....

IX.B.TUTOR RECYLING STRUCTURE .GRAPHICAL INTERPRETATION

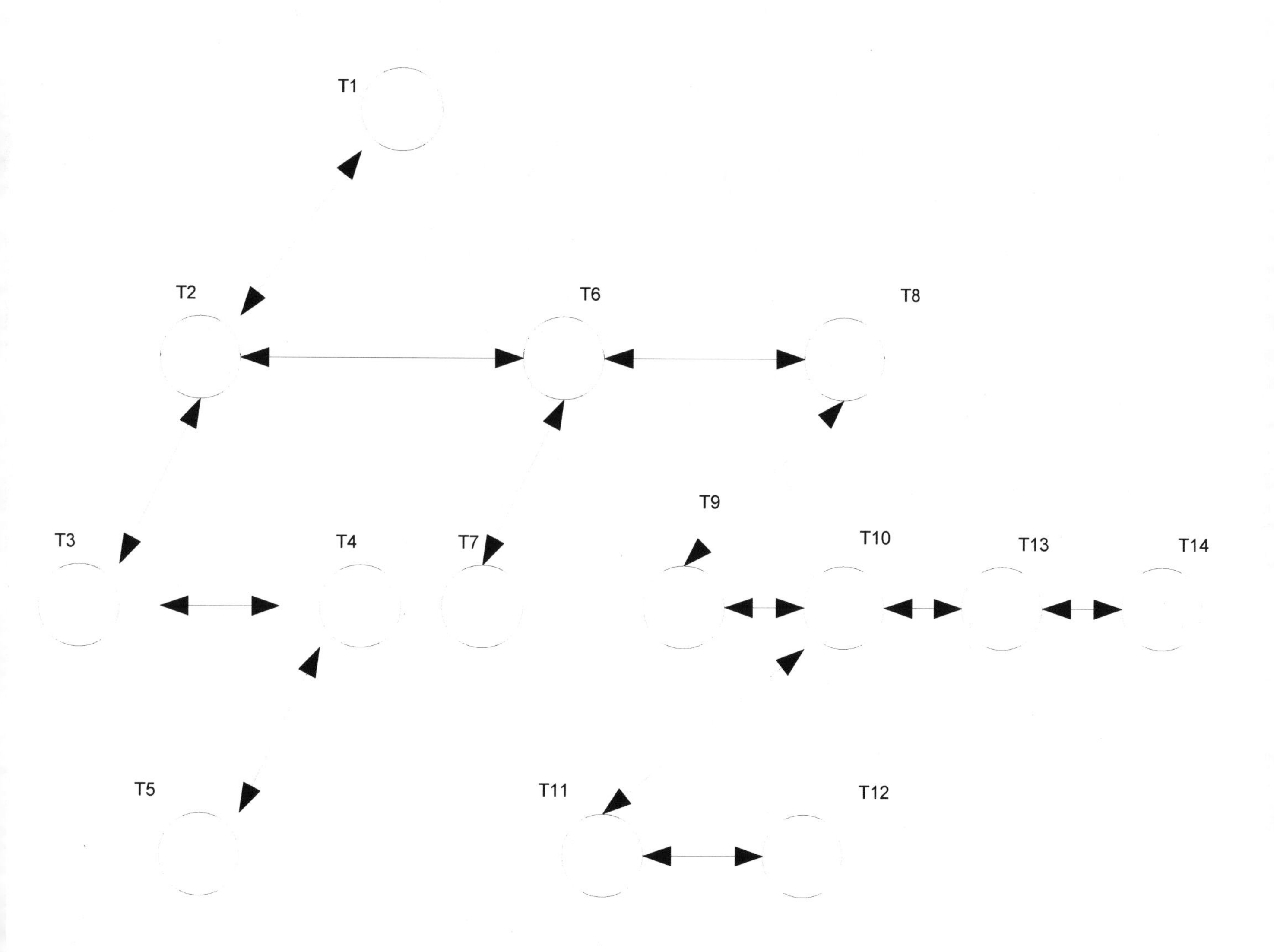

Msc.ing Feuzeu Yomsi
Raymond Jacquino

IX. RECYCLING AND NETWORKING

Because of the high demand of the tutoring in the environment, the OSEP will need to make a recycling and ensure a system of networking among the teachers in order to ensure good results .

A. Graphical Interpretation

B. Functional Description

The tutor number 1 shares and exchange ideas with tutor number 2

The tutor number 2 shares and exchange ideas with tutor number 6

Tutor number 8 and tutor number 3

Tutor number 3 shares and exchange ideas with tutor number 4 and tutor number 5

Tutor number 6 shares and exchange ideas with tutor number 8 and tutor number 7

Tutor number 8 shares and exchange ideas with tutor number 9

Tutor number 9 shares and exchange ideas with tutor number 10

Tutor number 10 shares and exchange ideas with tutor number 13 and tutor number 11

Tutor number 13 shares and exchange ideas with tutor number 1

Tutor number 11 shares and exchange ideas with tutor number 12

VII .OSEP OFFICE LOCATION STRATEGY

In attempt to satisfy all the students across America the OBAMA STIMULUS EDUCATION PLAN will be located

1.Office at every University across the country

2. Office inside every library across the country

3.independent office across the country .

4. etc.......

VIII THE OSEP OPPORTUNITY

. low fee to be OSEP member

.Training at the office , home on line

.flexible hours consultation

.Recruiting Tutors all category

. Get info on line or at the OSEP office

.etc......

C. Philosophy of the OBAMA STIMULUS EDUCATION PLAN

1.minimize the difficulties faced by all students .Encourage students to share experience between them. We expected/ hoped also student to become tutors.

2.Recycling or networking of the tutors

The recycling of the tutors of the OBAMA STIMULUS EDUCATION PLAN would prepare tutors to respond to the student demand . Tutors will challenge USA education system . The OSEP will be a resource of researchers in this country.

The productivity of the brain of the OBAMA STIMULUS EDUCATION PLAN will be in demand across the country :in the scientific area , Nasa, Automakers for new environment -friendly models, industrial sector ,.....

X. EXPECTATION FOR STUDENTS
A.GRAPHICAL INTERPRETATION

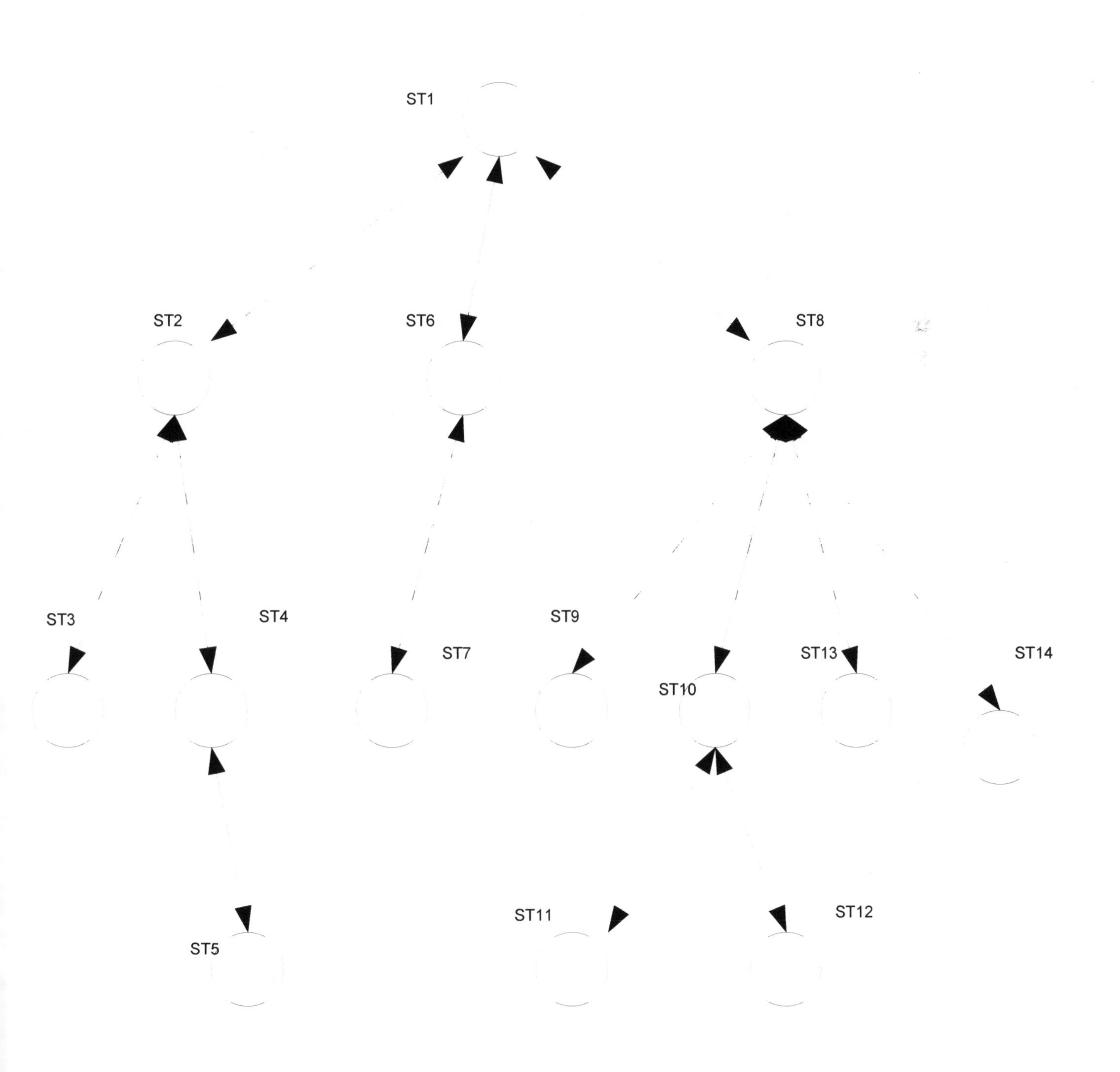

Msc.ing Feuzeu Yomsi
Raymond Jacquino

X EXPECTATION FOR STUDENTS

A. Graphical Interpretation

B. Description functional

One of the expectation for every student member of the OBAMA STIMULUS EDUCATION PLAN is that : every student should share the knowledge he/she receive from this plan with others students in needed , also he/she become a tutor

Student number 1 shares knowledge with student number 2, student number 6 and student number 8.
Student number 2 shares knowledge with student number 3 and student number 4

Student number 4 shares knowledge with student number 5

Student number 6 shares knowledge with student number 7

Student number 8 shares knowledge with student number 9
Student number 10 shares knowledge with student number 13 and with student number 14

Student 10 shares knowledge with student 11 and with student 12

OBAMA STIMULUS EDUCATION PLAN INFORMATION

I. STUDENT IDENTIFICATION

1.Name

2.Address

3.State

4.city

5.zip code

6.Social Security Number(last digit)

7.telephone number

8.email

9.Education grade

10.Name of the school attend now:

address:____

____

____

11. Available time for consultation____

12. Enumerate all academic difficulties face and send a copy to the OSEP office via email ____

13. Indicate the efforts attempts made to resolve the enumerate problem and send to OSEP administration a copy to ______________________________________

14. type of service need a)or b):____

a) Consultation at the office number______________________________

b)Consultation on line with password from OSEP administration

15. how many hours of consultation your need ?__________________________________

II. INFORMATION PROCESSING STEPS

step1 Student enumerate all academic difficulties he/she faces and contacts the OBAMA STIMULUS EDUCATION PLAN office .
The student will also encouraged to indicate the efforts and attempts made to resolve the enumerated problems.

Step2 The OBAMA STIMULUS EDUCATION PLAN office will immediately identifies and contact the appropriate tutor committee to see up an appointment .

Step3 The OBAMA STIMULUS EDUCATION PLAN tutors committee will provide the appropriate support to the student in need.

III. TREATMENT

1. begin
2. student data
3.indicate the efforts and attempts made
4.contact with the tutor committee appropriate

<u>if</u> the efforts and attempts made by the student is approved
by the appropriate tutor <u>then</u> he provide necessary support
<u>else</u> the student need to increase or make extra efforts and attempts.

i. CHOISE OF THE IMITATION CLASS
code k =1, k=2, k= 3

1. End of the day at the OBAMA STIMULUS EDUCATION PLAN STATION OR OFFICE

2. REGENERATION : THE STUDENT S RECEIVE THE APPROPRIATE SUPPORT HE NEED

3. End of the day :collect of the statistics informations

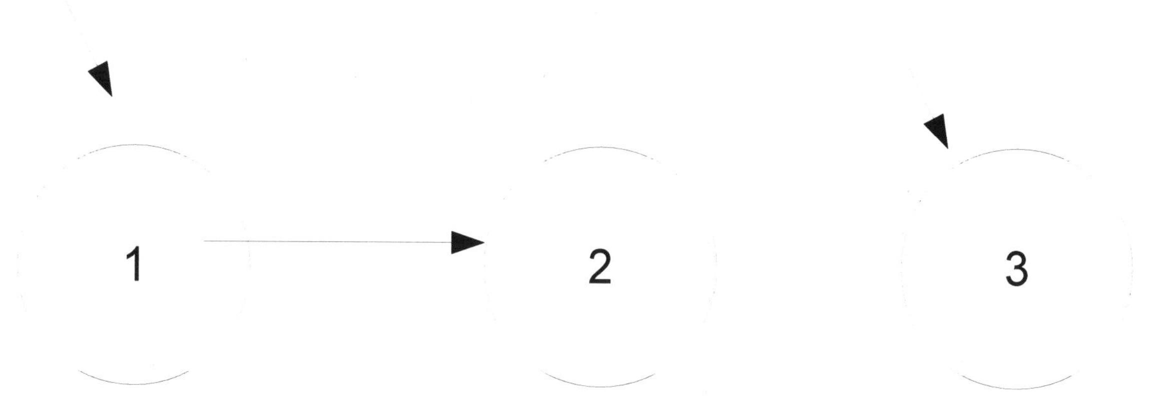

(1-2) The student is not able to solve by himself the academic difficulties he/she faces need support

Feuzeu Yomsi
Raymond Jacquino

2.ALGORITHME
TREATMENT OF THE EVENT K=1

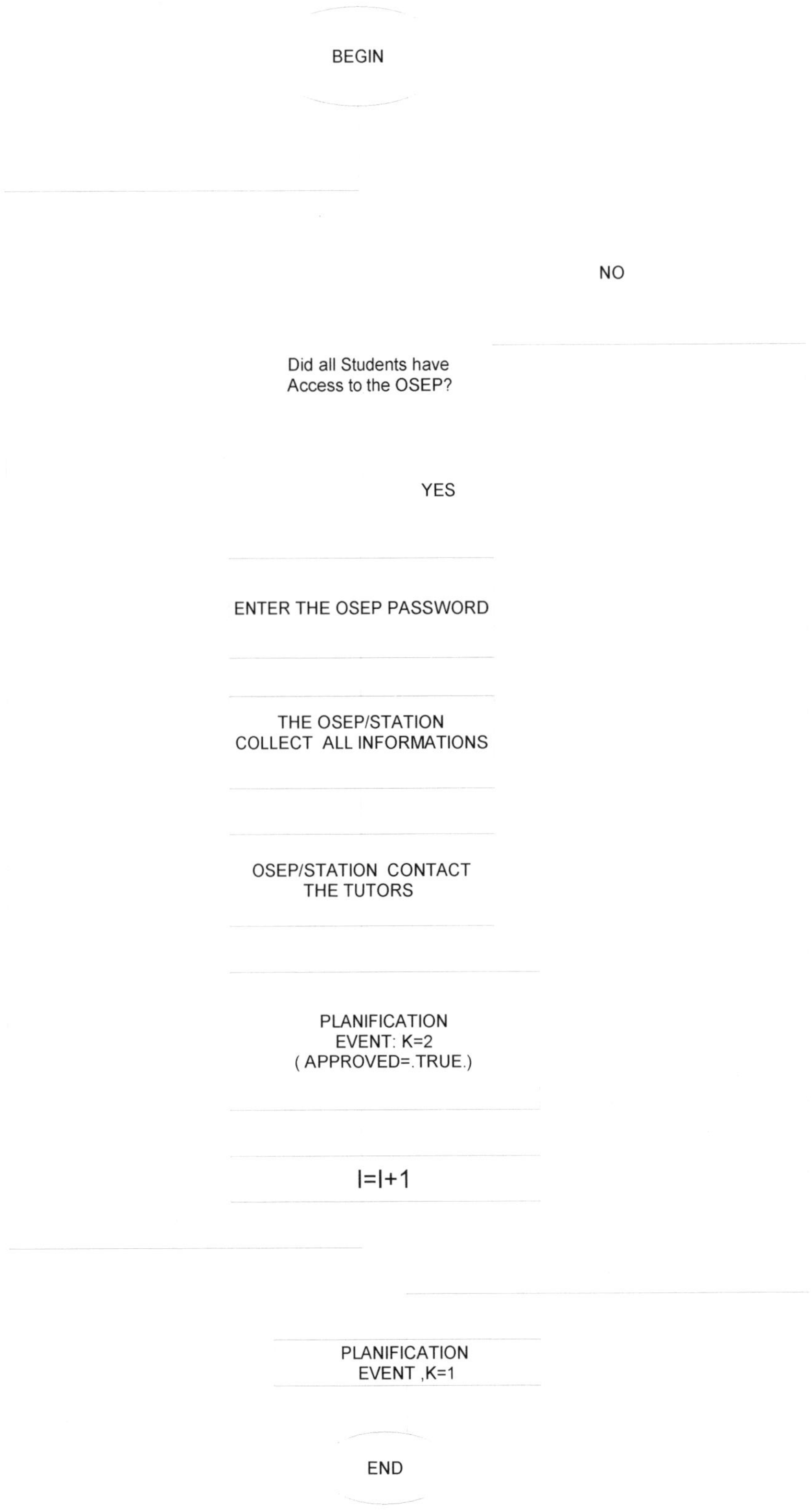

BEGIN

NO

Did all Students have
Access to the OSEP?

YES

ENTER THE OSEP PASSWORD

THE OSEP/STATION
COLLECT ALL INFORMATIONS

OSEP/STATION CONTACT
THE TUTORS

PLANIFICATION
EVENT: K=2
(APPROVED=.TRUE.)

I=I+1

PLANIFICATION
EVENT ,K=1

END

Feuzeu Yomsi
Raymond Jacquino

ii. ALGORIGTHME
TREATMENT EVENT CODE K=2

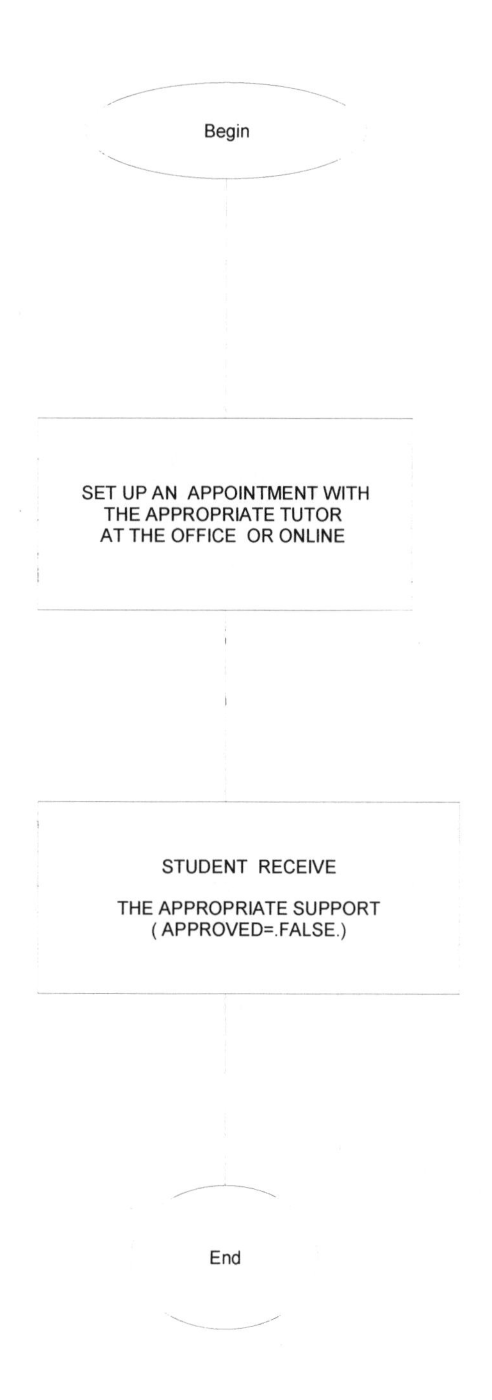

Feuzeu Yomsi
Raymond Jacquino

ii. ALGORIGTHME
TREATMENT EVENT CODE K=3

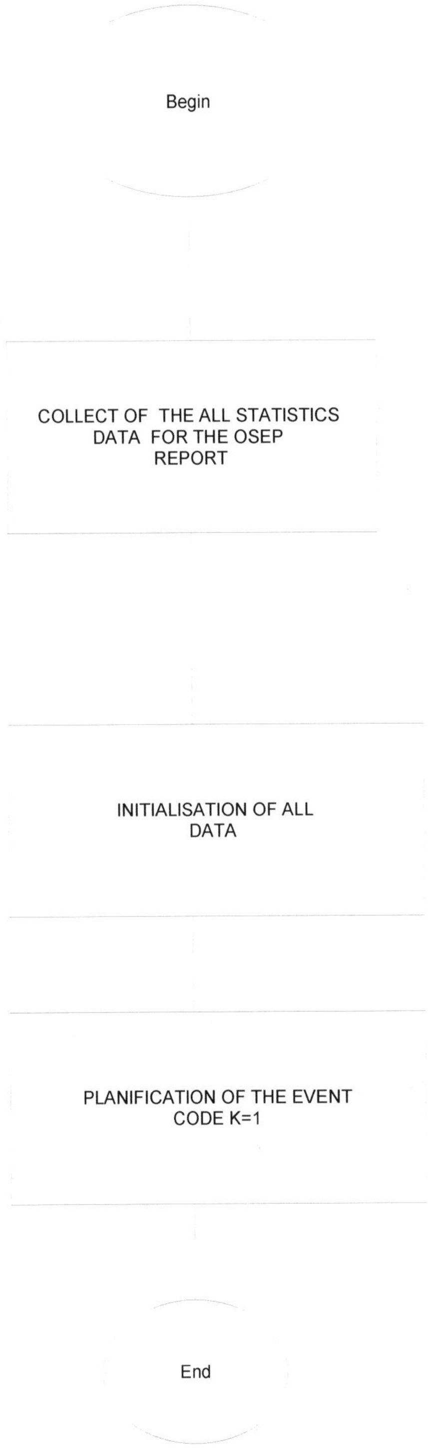

Begin

COLLECT OF THE ALL STATISTICS
DATA FOR THE OSEP
REPORT

INITIALISATION OF ALL
DATA

PLANIFICATION OF THE EVENT
CODE K=1

End

Feuzeu Yomsi
Raymond Jacquino

XI. TEST AND ANALYSE OF THE OSEP

A. Entry Data

a) number of state =2 :PA and OHIO

1. **state: Pa**
   city : Philadelphia(PHL)
2. **state :OHIO**
   city : Chicago

b) number of OBAMA STIMULUS EDUCATION office in PA and OHIO=5

OSEP/pa/office1=id1=100a
OSEP/pa/office2=id2=101a
OSEP/OHIO/office1=id3=200b
OSEP/OHIO/office2=id4=201b
OSEP/OHIO/office3=id5=202b

c)number of students who receive a consultation at OBAMA stimulus education plan office in Pa and in OHIO= 10

d)students receive appropriate consultation in :

1.mathematic in Philadelphia
2.Information technology in Chicago

B. The central administration in Washington need the situational report of the OSEP across America in OHIO and Pennsylvania in this particular case .
Eventually needed to know the average of satisfy students or not satisfy by the OSEP.

Our goal is to minimize the academy difficulties face by the students, maximize the success , increase the level of knowledge .

C. Graphical Interpretation

XI.TEST AND ANALYSE OF THE OSEP EFFECT

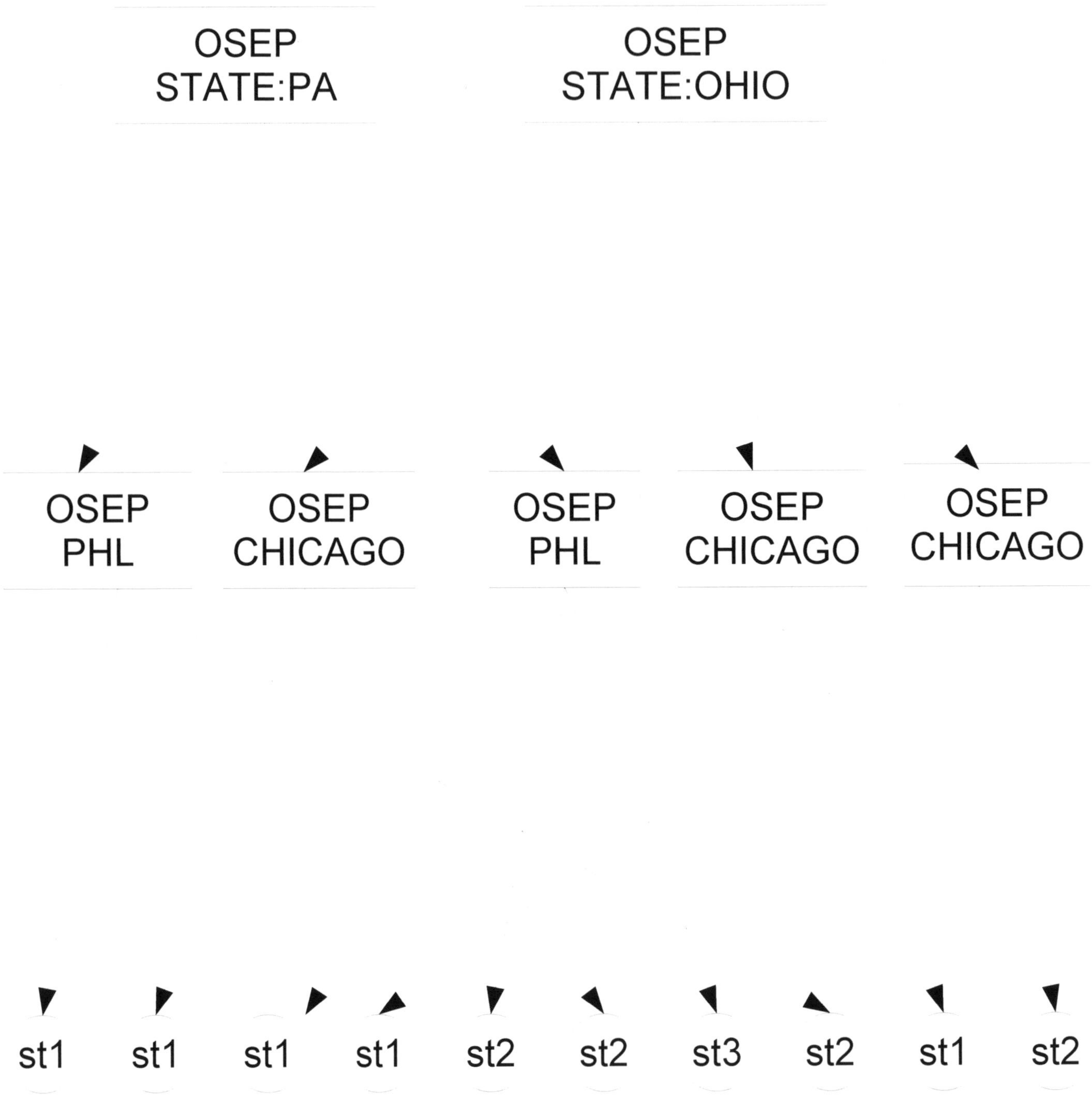

OSEP
STATE:PA

OSEP
STATE:OHIO

OSEP
PHL

OSEP
CHICAGO

OSEP
PHL

OSEP
CHICAGO

OSEP
CHICAGO

st1 st1 st1 st1 st2 st2 st3 st2 st1 st2

The OSEP central control Need the situational report of every OSEP office in the Pennsylvania and OHIO.

Msc.ing Feuzeu Yomsi
Raymond Jacquino

X1.C.ANALYSE OF OBAMA STIMULUS EDUCATION PLAN EFFECT IN 4 YEARS in USA

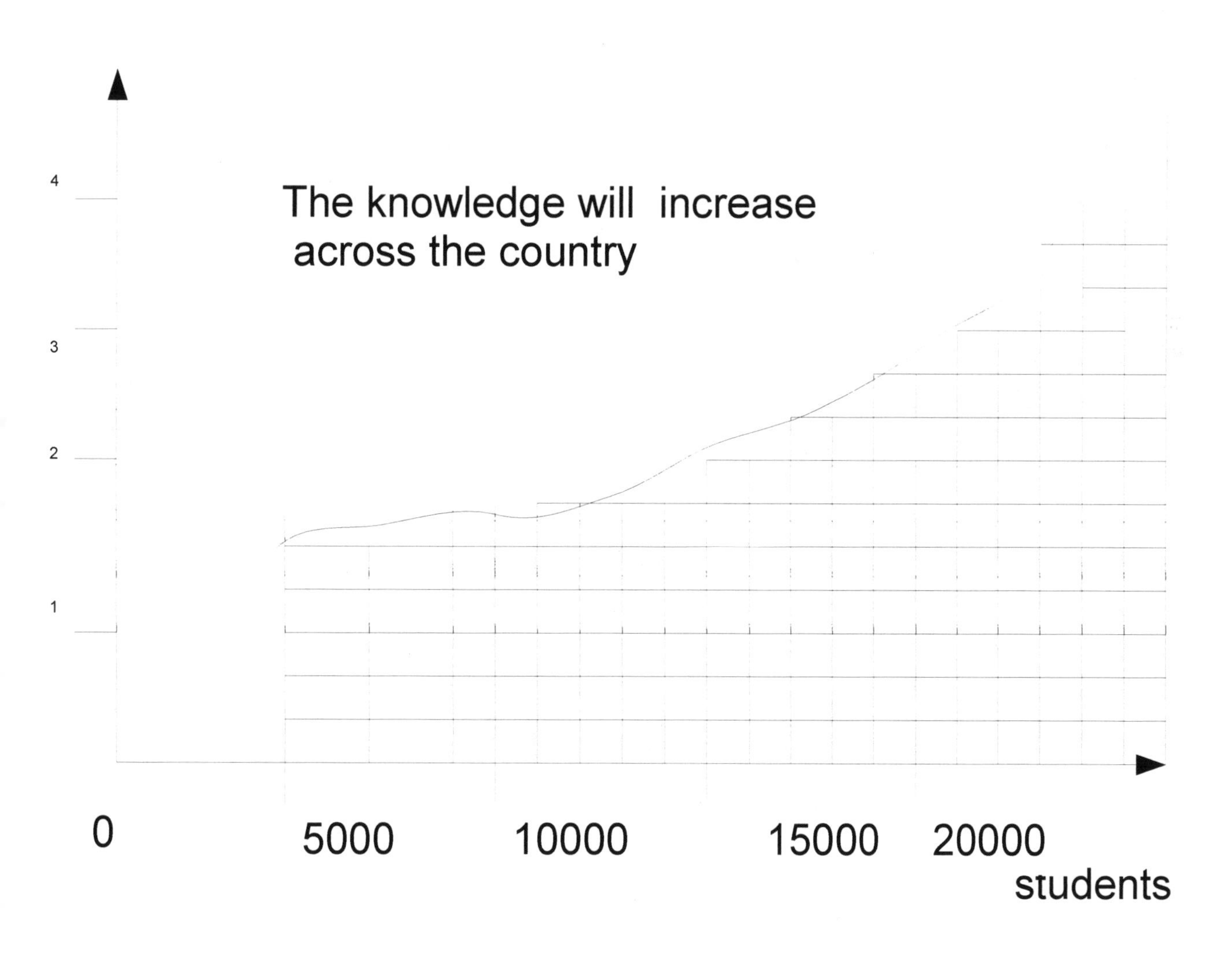

Msc.ing Feuzeu Yomsi
Raymond Jacquino

X . FURNITURE INVENTORY

1.office furniture

2. OSEP software

3.computer

4.printer

5.photocopier

6. etc....

XI.THE COST OF THE OSEP

XII . CONCLUSION.

APPRECIATION

Msc.-ing Feuzeu Yomsi Raymond Jacquino master in information processing control is the author of this document dedicate to our president BARACK OBAMA .
America is the place where anything- anything we choose to dream together, anything for which we choose to work together --is possible.
This is our time even as we celebrate, we know the challenges are the greatest of our lifetime . The OBAMA expectation is very high that is the reason why I designed here a stimulus education package that can " jolt the education into shape ". The plan will create millions of jobs , challenge our education system, motivate our students , help thousands of students to become successful in all academic areas by putting in place a network of tutoring to assist them.

This victory alone is not the change we seek. It is the only chance for me to contribute for the OBAMA change and I believe this education plan will help people back to work , open door to our kids , restore prosperity and the progress of United state of AMERICA .

IN GOD WE TRUST

KENYA REVOLUTION

2009

EDUCATION IS POWER

OBAMA

STIMULUS EDUCATION PLAN (OSEP)
FOR THE LAND OF MY FATHER'S BIRTH

THE VOICE OF THE AUTHOR
Msc.-ing Feuzeu Yomsi
Raymond Jacquino

I. AUTOMATIZATION OF THE OBAMA STIMULUS EDUCATION PLAN

A.FUNCTIONAL STRUCTURE OF THE SYSTEM OSEP

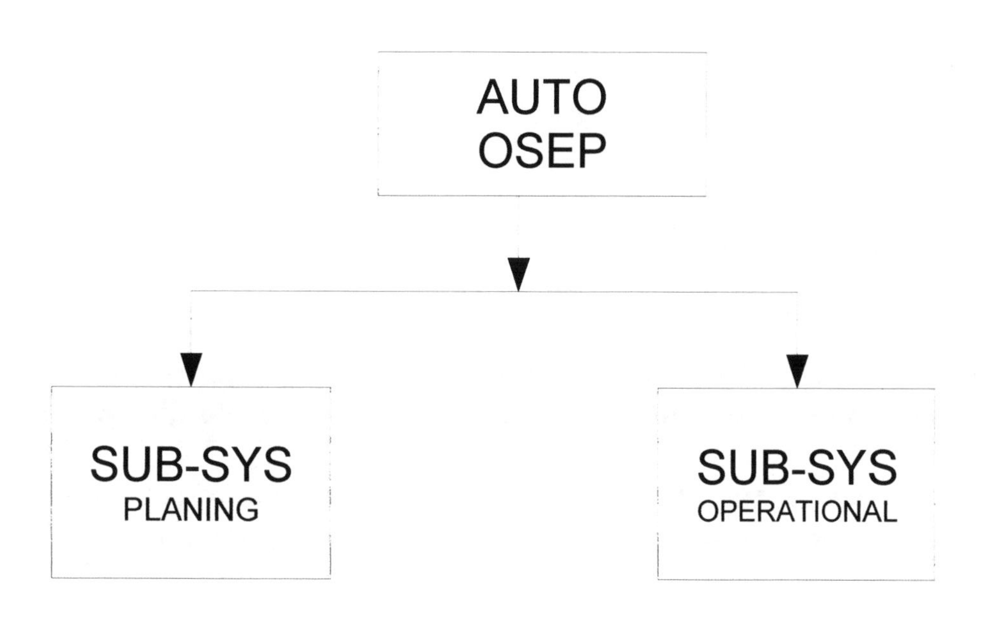

B.RELATIONAL DIAGRAM

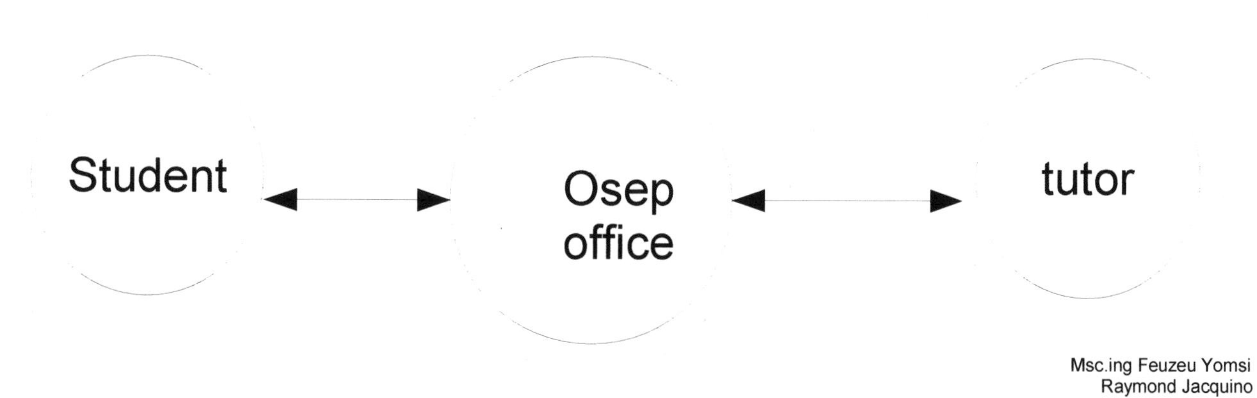

Msc.ing Feuzeu Yomsi
Raymond Jacquino

I.AUTOMATIZATION OF THE OBAMA STIMULUS EDUCATION PLAN (OSEP)

a. Introduction

The key features of this automatization system include two majors systems. sub-systems: sub-system planing and operational sub-system. we will focus on the sub-system planing : that means the OPTIMIZATION OF THE OBAMA EDUCATION PLAN.

b. OPTIMIZATION OF THE OBAMA STIMULUS EDUCATION PLAN

The key features of the program include:

Motivation: OSEP will motivate all students across the country to become successful in all academic areas of the curriculum.

Assistance: Because of the high demand of tutoring in the environment OSEP will create a network of assistance to students

Minimization: Our goal is to minimize the academy difficulties face by students in school

Maximization: Eventually maximizing the success , increasing the level of knowledge, etc ...

c. The relation between a student and the tutor via the administration of the OBAMA stimulus education plan

VI. A.STUDENT CONSULTATION AND APPRECIATIONOF THE COMMITTEE

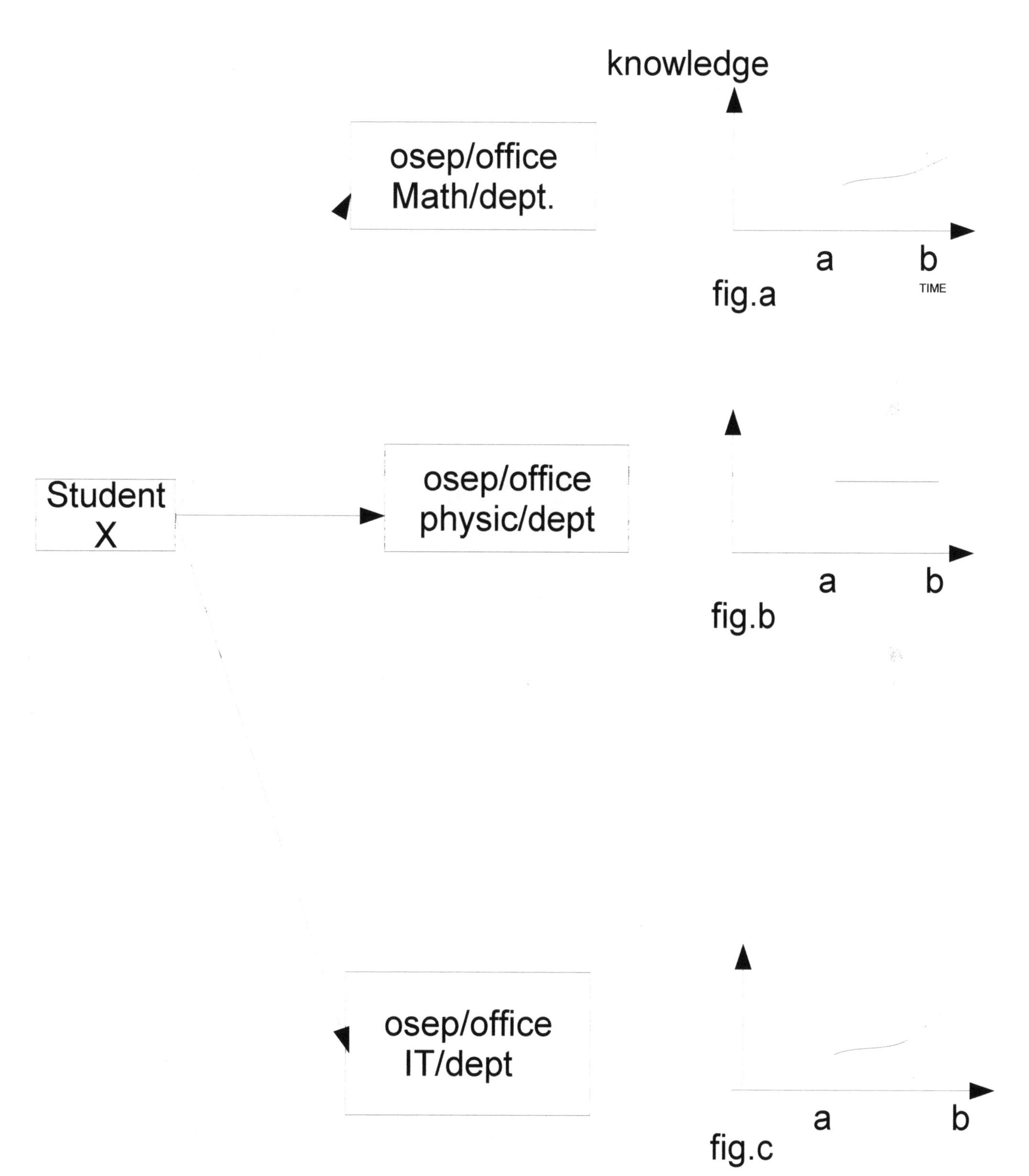

Msing. Feuzeu Yomsi
Raymond Jacquino

VI . STUDENT CONSULTATION AND APPRECIATION OF THE COMMITTEE

a. Graphical Interpretation

b. Description

Student x	**request assistance in mathematic and obtained**	**fig a. Good**
Student x	**request assistance in physic and obtained**	**fig b. Constant**
Student x	**request assistance in IT.**	**fig c. excellent**

VI I. PARAMETERS OF THE SITUATIONAL REPORT OF EVERY OFFICE

IDOSEP:Identification number of the OBAMA STIMULUS EDUCATION PLAN office

STAOSEP:State where located

CTY: Name of the city

NOSEP: Number of the students who receive a consultation at OSEP office a day,a week, year

SOSEP: Summary of students satisfy with this OSEP at that office

SnOSEP:Summary of students not satisfy with this OSEP

Ntmath:Number of tutors in mathematic committee of the OSEP in that office

Ntphy:Number of tutors in physic committee of the OSEP in that office

NtIT:Number of tutors in Information Technology committee of the OSEP in that office

etc....

MsucOSEP:The average of the students who became successful with the OBAMA STIMULUS EDUCATION PLAN in that office

MnsucOSEP:The average of the students not satisfy or unsuccessful with the OSEP

Ti: Total number of hours of consultation offered to student , etc ....

V. THE OSEP TUTORS SELECTION AND FUNCTIONAL DESCRIPTION

A. RECRUITMENT OF CVTUTORS

The OSEP will make announcement via Internet , TV,Radio,Newspapers.

The OSEP application will be on line or at the OSEP office available with all conditions request and the OSEP selection will follow up to see who is better qualify for the job

B. FUNCTIONAL DESCRIPTION OF THE OSEP

1. **Tutors of the mathematic committee of the OBAMA STIMULUS EDUCATION PLAN will help students member of the OSEP become successful in mathematic area by offering solutions their academic difficulties.**

2. **Tutors of the physics committee of the OBAMA STIMULUS EDUCATION PLAN will help student members solutions their academic difficulties in physic**

..

n. Tutors of the Informations Technology committee will help students become successful by minimize in Information Technology difficulties face by them .

3. The service will be at one of the OSEP office or on line

II. INFORMATION PROCESSING STEPS

step0 Student enumerate all academic difficulties he/she faces and contacts the OBAMA STIMULUS EDUCATION PLAN office. The student will also be encouraged to indicate the efforts and attempts made to resolve the enumerated problems.

step1 The OBAMA STIMULUS EDUCATION PLAN office immediately Identifies and contact the appropriate tutor committee to set up an appointment

step2 The OBAMA STIMULUS EDUCATION PLAN tutors provide the appropriate support to the student in need .

II .PARAMETER DESIGNATION

The OSEP will constitute or assemble a network of tutors sub divided into various key disciplines(mathematics, physics,... information technology, etc.) that is mathematic tutors committee , a physic tutors committee, information technology tutors committee etc...

IV STUDENTS NEEDS AND RESPONSIBILITIES

1. Registered with OSEP , to be OSEP member
2. Study to become successful
3. Identify or enumerate he/her difficulties
4. Give evidence of personal attempts to resolve problem
5. Be ready eventually to become an OSEP tutor to help others in need
6. etc....

IX.B.TUTOR RECYLING STRUCTURE .GRAPHICAL INTERPRETATION

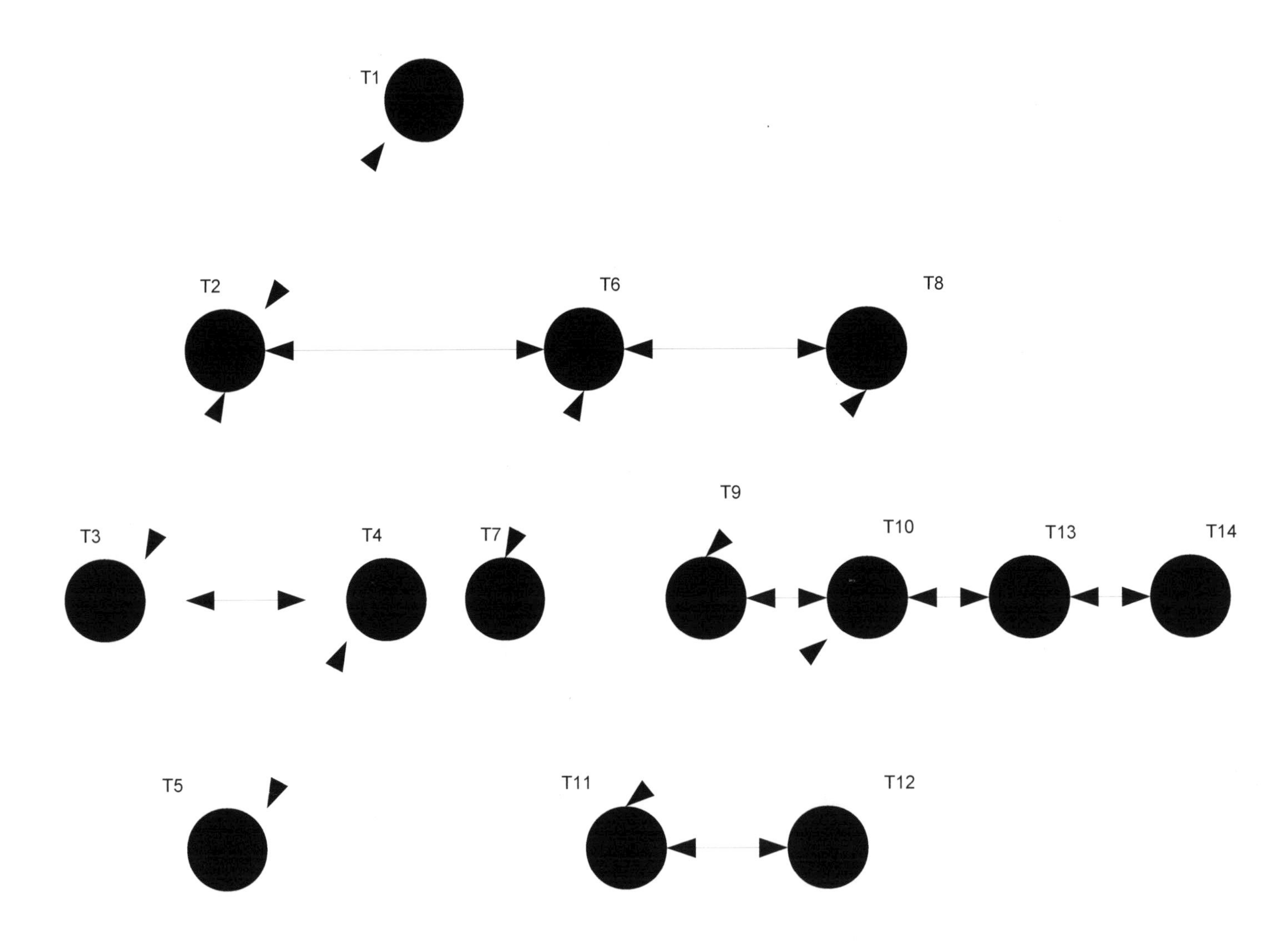

Msc.ing Feuzeu Yomsi
Raymond Jacquino

IX. RECYCLING AND NETWORKING

Because of the high demand of the tutoring in the environment, the OSEP will need to make a recycling and ensure a system of networking among the teachers in order to ensure good results .

A. Graphical Interpretation

B. Functional Description

The tutor number 1 shares and exchange ideas with tutor number 2
The tutor number 2 shares and exchange ideas with tutor number 6

Tutor number 8 and tutor number 3
Tutor number 3 shares and exchange ideas with tutor number 4 and tutor number 5

Tutor number 6 shares and exchange ideas with tutor number 8 and tutor number 7

Tutor number 8 shares and exchange ideas with tutor number 9

Tutor number 9 shares and exchange ideas with tutor number 10

Tutor number 10 shares and exchange ideas with tutor number 13 and tutor number 11

Tutor number 13 shares and exchange ideas with tutor number 1

Tutor number 11 shares and exchange ideas with tutor number 12

VII .OSEP OFFICE LOCATION STRATEGY

In attempt to satisfy all the students across America the OBAMA STIMULUS EDUCATION PLAN will be located

1.Office at every University across the country

2. Office inside every library across the country

3.independent office across the country .

4. etc.......

VIII THE OSEP OPPORTUNITY

. low fee to be OSEP member

.Training at the office , home on line

.flexible hours consultation

.Recruiting Tutors all category

. Get info on line or at the OSEP office

.etc......

C. Philosophy of the OBAMA STIMULUS EDUCATION PLAN

1.minimize the difficulties faced by all students .Encourage students to share experience between them. We expected/ hoped also student to become tutors.

2.Recycling or networking of the tutors

The recycling of the tutors of the OBAMA STIMULUS EDUCATION PLAN would prepare tutors to respond to the student demand . Tutors will challenge USA education system . The OSEP will be a resource of researchers in this country.

The productivity of the brain of the OBAMA STIMULUS EDUCATION PLAN will be in demand across the country :in the scientific area , Nasa, Automakers for new environment -friendly models, industrial sector ,.....

IX.B.TUTOR RECYLING STRUCTURE .GRAPHICAL INTERPRETATION

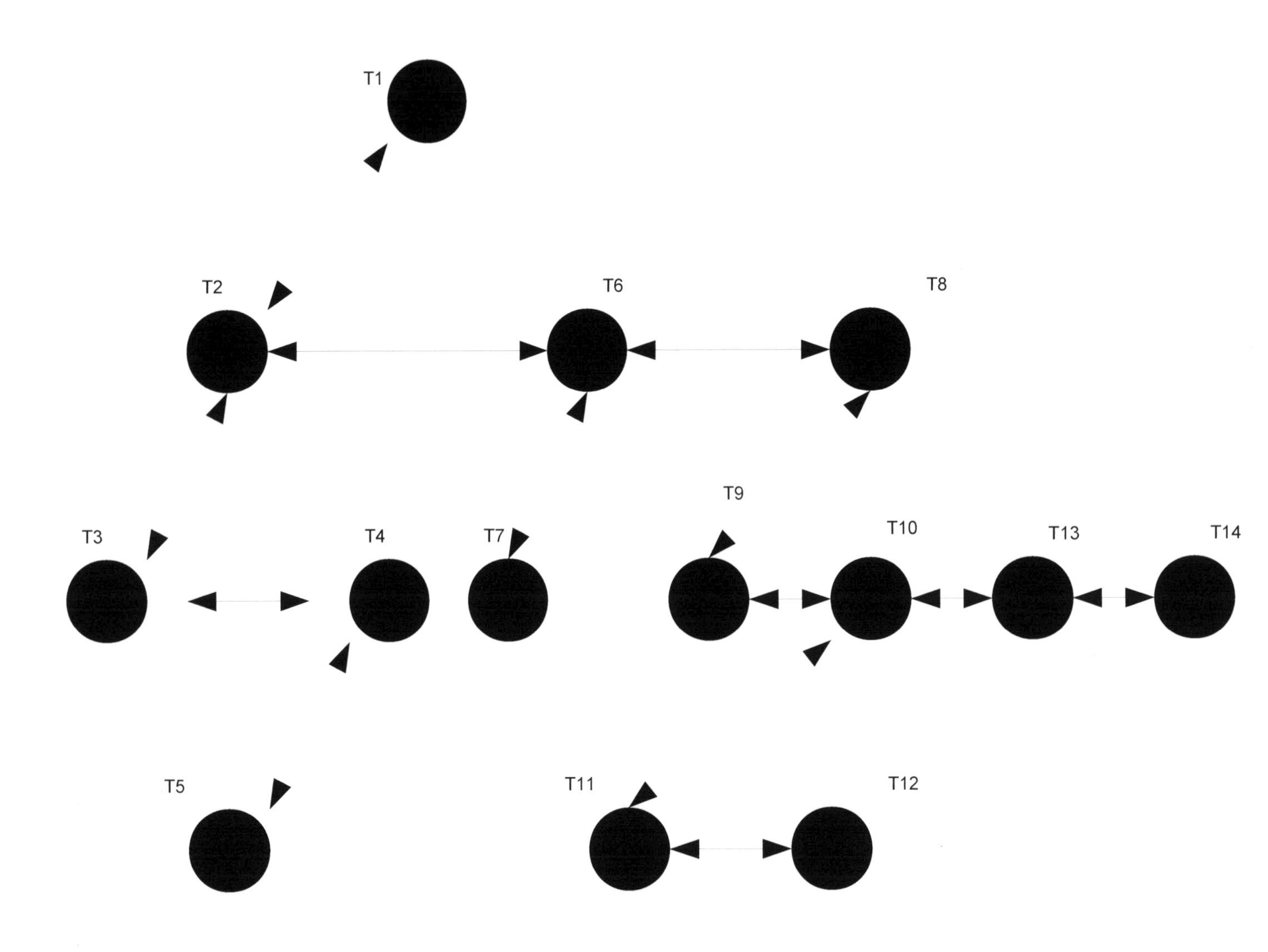

Msc.ing Feuzeu Yomsi
Raymond Jacquino

X EXPECTATION FOR STUDENTS

A. Graphical Interpretation

B. Description functional

One of the expectation for every student member of the OBAMA STIMULUS EDUCATION PLAN is that : every student should share the knowledge he/she receive from this plan with others students in needed , also he/she become a tutor

Student number 1 shares knowledge with student number 2, student number 6 and student number 8.
Student number 2 shares knowledge with student number 3 and student number 4

Student number 4 shares knowledge with student number 5

Student number 6 shares knowledge with student number 7

Student number 8 shares knowledge with student number 9
Student number 10 shares knowledge with student number 13 and with student number 14

Student 10 shares knowledge with student 11 and with student 12

OBAMA STIMULUS EDUCATION PLAN INFORMATION

I. **STUDENT IDENTIFICATION**

1.Name

2.Address

3.State

4.city

5.zip code

6.Social Security Number(last digit)

7.telephone number

8.email

9.Education grade

10.Name of the school attend now:

address:____

____

____

11. Available time for consultation____

12. Enumerate all academic difficulties face and send a copy to the OSEP office via email ____

13. Indicate the efforts attempts made to resolve the enumerate problem and send to OSEP administration a copy to ____

14. type of service need a)or b):____

a) Consultation at the office number______________________________

b)Consultation on line with password from OSEP administration

15. how many hours of consultation your need ?______________________________

II. INFORMATION PROCESSING STEPS

step1 Student enumerate all academic difficulties he/she faces and contacts the OBAMA STIMULUS EDUCATION PLAN office .
The student will also encouraged to indicate the efforts and attempts made to resolve the enumerated problems.

Step2 The OBAMA STIMULUS EDUCATION PLAN office will immediately identifies and contact the appropriate tutor committee to see up an appointment .

Step3 The OBAMA STIMULUS EDUCATION PLAN tutors committee will provide the appropriate support to the student in need.

III. TREATMENT

1. begin
2. student data
3.indicate the efforts and attempts made
4.contact with the tutor committee appropriate

<u>if</u> the efforts and attempts made by the student is approved
by the appropriate tutor <u>then</u> he provide necessary support
<u>else</u> the student need to increase or make extra efforts and attempts.

i. CHOISE OF THE IMITATION CLASS
code k =1, k=2, k= 3

1. End of the day at the OBAMA STIMULUS EDUCATION PLAN STATION OR OFFICE

2. REGENERATION : THE STUDENT S RECEIVE THE APPROPRIATE SUPPORT HE NEED

3. End of the day :collect of the statistics informations

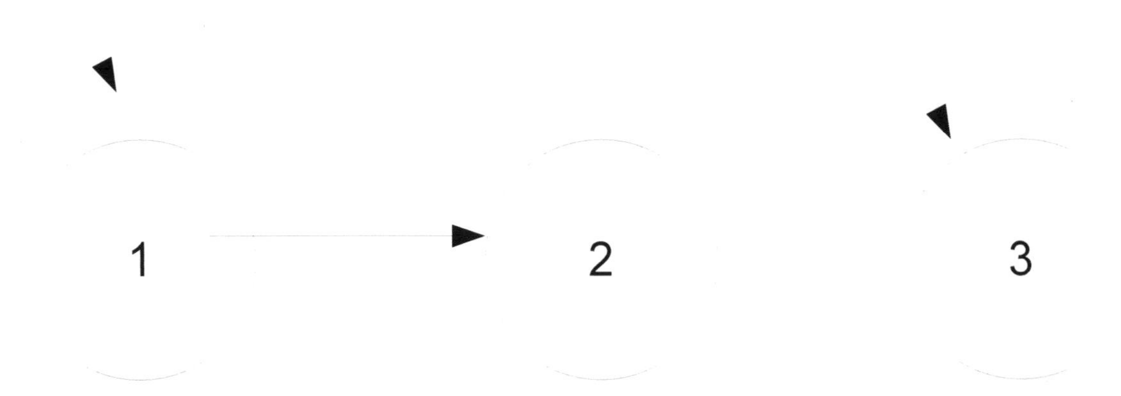

(1-2) The student is not able to solve by himself the academic difficulties he/she faces need support

Feuzeu Yomsi
Raymond Jacquino

2.ALGORITHME
TREATMENT OF THE EVENT K=1

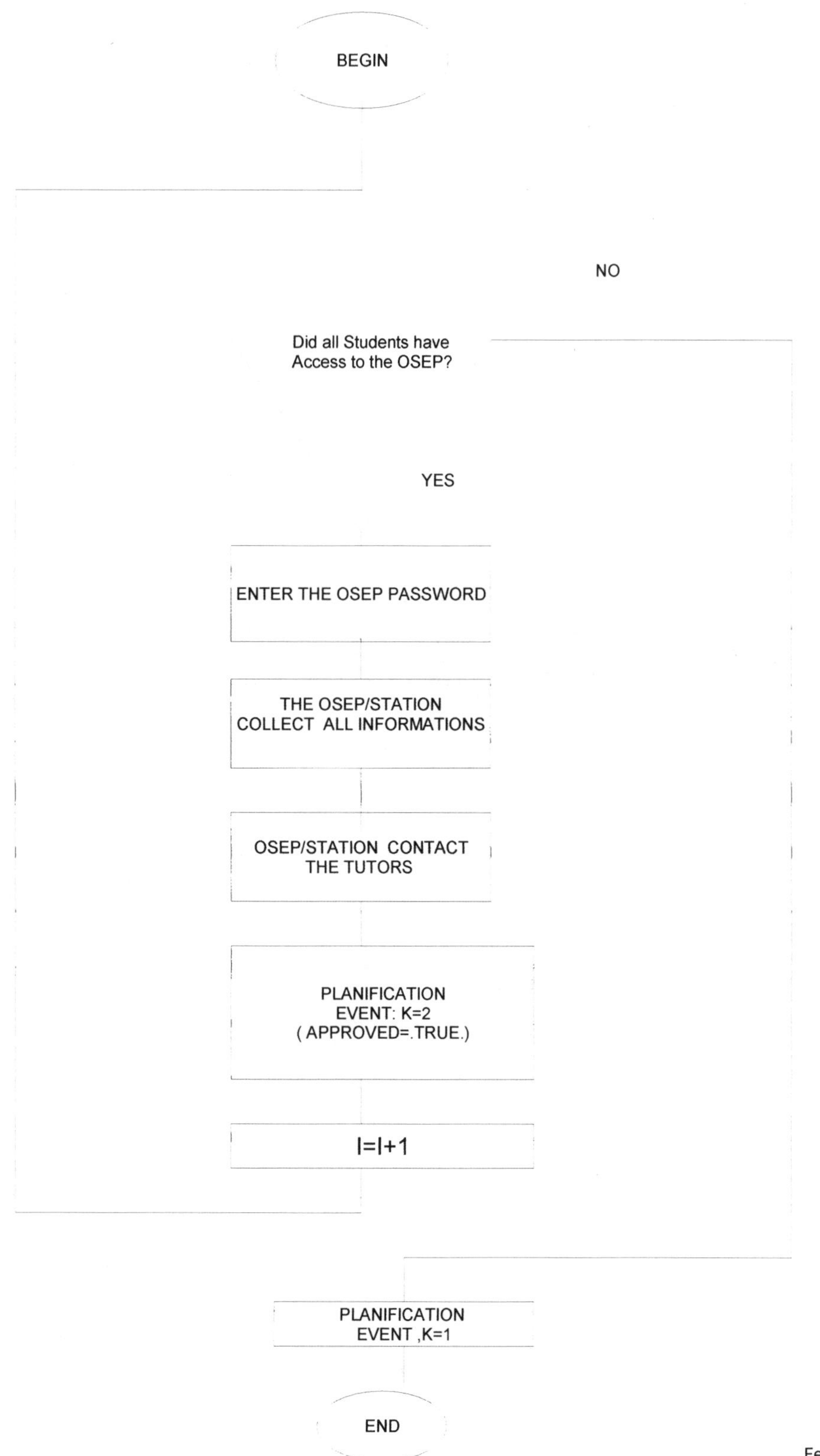

Feuzeu Yomsi
Raymond Jacquino

ii. ALGORIGTHME
TREATMENT EVENT CODE K=2

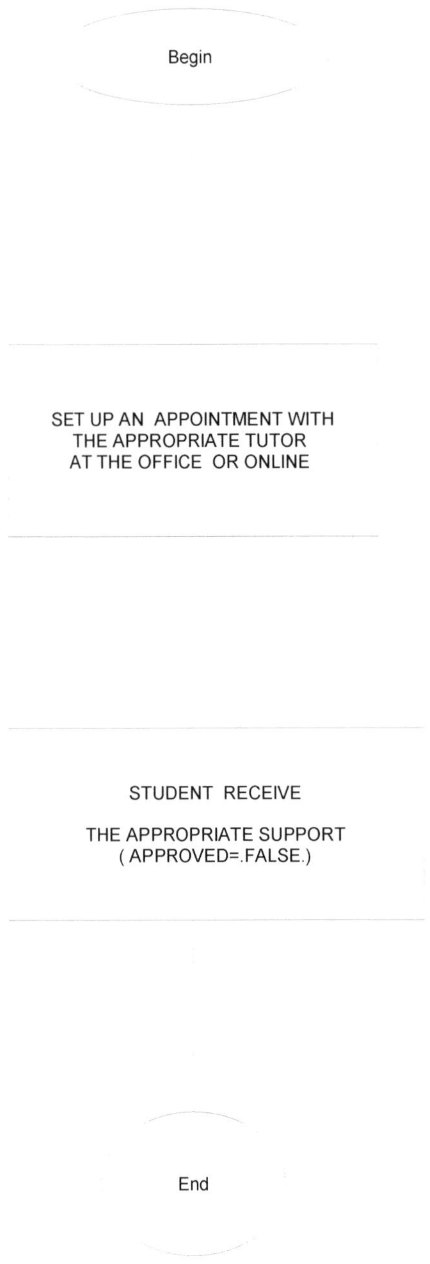

Begin

SET UP AN APPOINTMENT WITH
THE APPROPRIATE TUTOR
AT THE OFFICE OR ONLINE

STUDENT RECEIVE

THE APPROPRIATE SUPPORT
(APPROVED=.FALSE.)

End

Feuzeu Yomsi
Raymond Jacquino

ii. ALGORIGTHME
TREATMENT EVENT CODE K=3

Begin

COLLECT OF THE ALL STATISTICS DATA FOR THE OSEP REPORT

INITIALISATION OF ALL DATA

PLANIFICATION OF THE EVENT CODE K=1

End

Feuzeu Yomsi
Raymond Jacquino

XI .TEST AND ANALYSE OF THE OBAMA STIMULUS EDUCATION PLAN(OSEP)

A. Entry Data:

a)number of cities in Kenya

1. **Kenya**
   city :Nariobi
2. **Kenya**
   city:kisumu

b)Number of OBAMA stimulus office or station in Nariobi and Kisumu=5

OSEP/Kenya/Nariobi/office1=id1=100b
OSEP/Kenya/Nariobi/office2=id2=101b
OSEP/Kenya/Kisumu/office1=id3=201c
OSEP/Kenya/Kisumu/office2=id4=202d
OSEP/Kenya/Kisumu/office3=id5=203e

c)Number of students who receive a consultation at OBAMA education office in Nariobi and Kisumu =10

d) Students receive appropriate consultation in:

1. mathematic in Nariobi
Information technology in Kisumu

B. the central administration of the OBAMA Stimulus Education Plan need the situational report across Kenya in particular case in Nariobi and Kisumu . Eventually needed to know the average of satisfy students or not satisfy by the OBAMA Stimulus Education Plan .

Our goal is to minimize the academy difficulties face by students of Kenya , maximize the success , increase the level of knowledge.

c.Graphical Interpretation

XI.TEST AND ANALYSE OF THE OSEP EFFECT

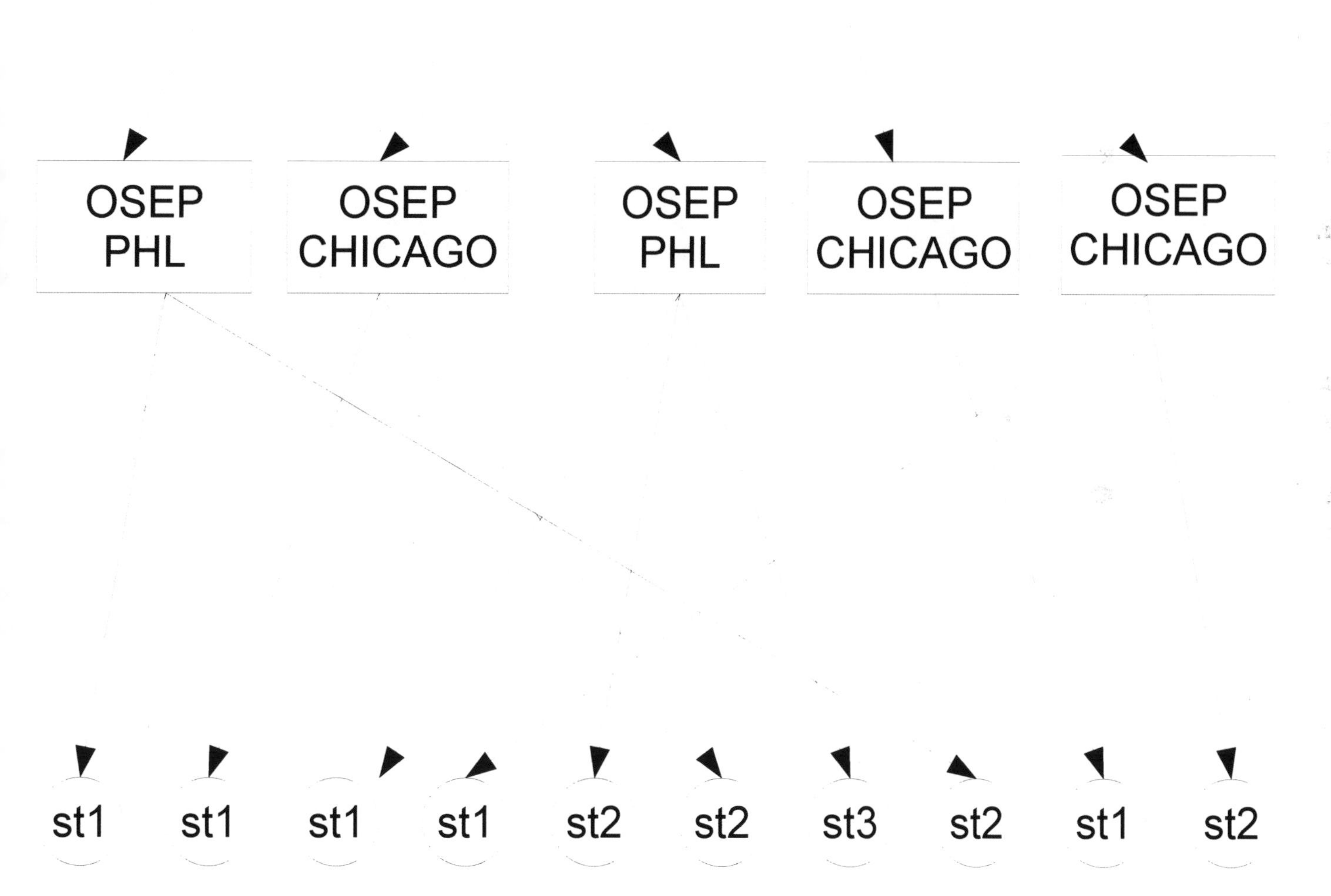

The OSEP central control Need the situational report of every OSEP office in the Pennsylvania and OHIO.

Msc.ing Feuzeu Yomsi
Raymond Jacquino

X1.C.ANALYSE OF OBAMA STIMULUS EDUCATION PLAN EFFECT IN 4 YEARS

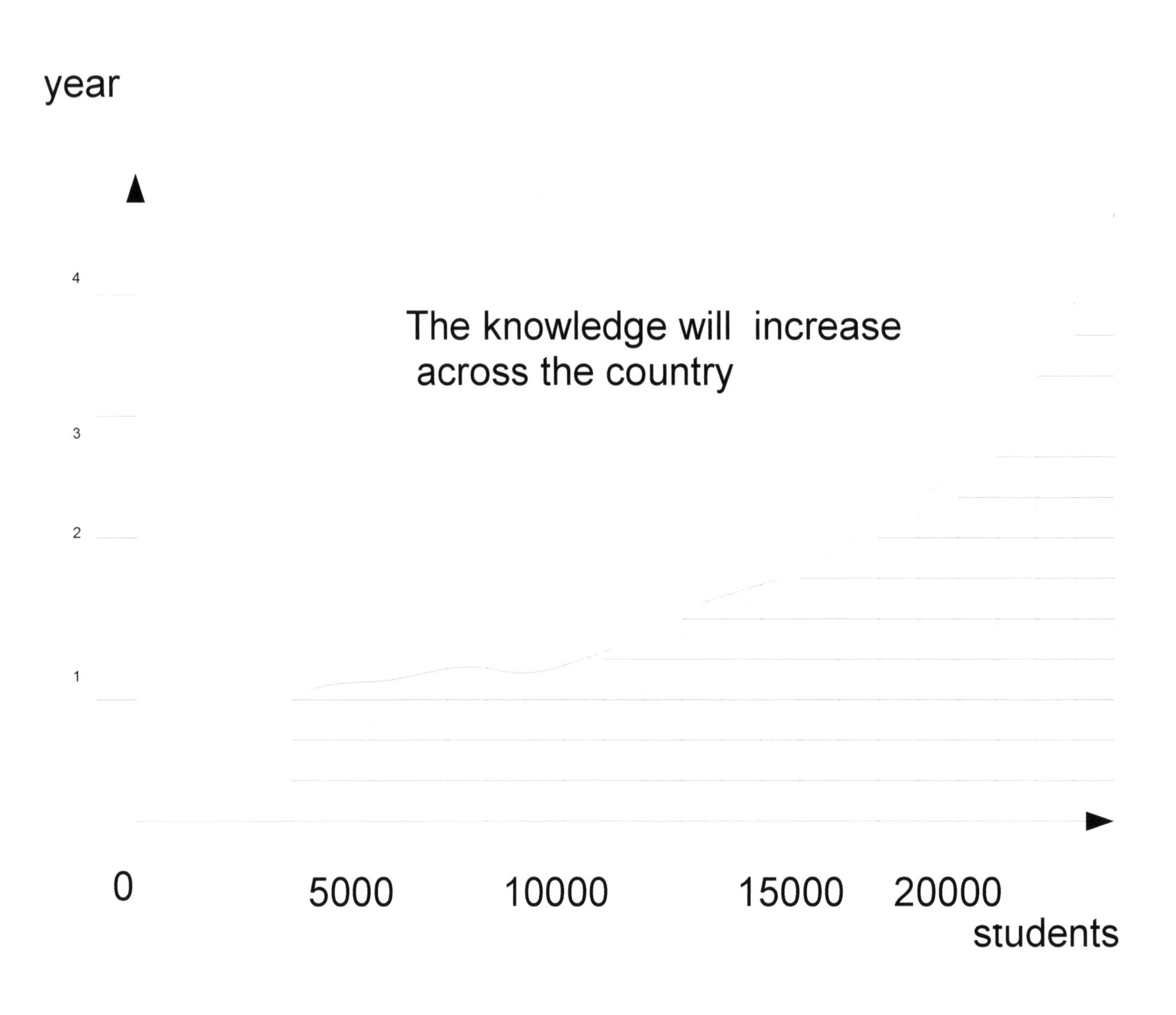

Msc.ing Feuzeu Yomsi
Raymond Jacquino

X . FURNITURE INVENTORY

1. office furniture
2. OSEP software
3. computer
4. printer
5. photocopier
6. etc....

XI.THE COST OF THE OSEP

XII . CONCLUSION.

APPRECIATION

Msc.-ing Feuzeu Yomsi Raymond Jacquino Master in Engineering in information processing control . Education is a power, I have the honor to design this book to help our president BARACK OBAMA . The plan if adopted will be named: OBAMA STIMULUS EDUCATION PLAN. The plan will challenge the education system, will help thousands of kenyan students to became successful in all academic areas by putting in place a Network of tutoring to assist students, the plan will open doors of opportunity to kids, restore properity and make Kenya progress.

\

Author:
Msc.-ing Feuzeu Yomsi
Raymond Jacquino

The president of the Cameroonian Association in Greater Philadelphia

IN GOD WE TRUST

XI. TEST AND ANALYSE OF THE OSEP

A. Entry Data

a) number of City =2 : Yaounde and Douala

1. **CAMEROON**
   city : Yaounde
2. **CAMEROON**
   City :Douala

b) number of OBAMA STIMULUS EDUCATION office in and OHIO=5

OSEP/Cameroon/Yaounde/office1=id1=100a
OSEP/Cameroon/Yaounde/office2=id2=101a
OSEP/Cameroon/Douala/office1=id3=200b
OSEP/Cameroon/Douala/office2=id4=201b
OSEP/Cameroon/Douala/office3=id5=202b

c)number of students who receive a consultation at OBAMA stimulus education plan office in Yaounde and in Douala= 10

d)students receive appropriate consultation in :

1.mathematic in Yaounde
2.Information technology in Douala

B. The central administration in Cameroon need the situational report of the OSEP across the country in this particular case in Yaounde and Douala. Eventually needed to know the average of satisfy students or not satisfy by the OSEP.

Our goal is to minimize the academy difficulties face by the students, maximize the success , increase the level of knowledge .

C. Graphical Interpretation

XI.TEST AND ANALYSE OF THE OSEP EFFECT In CAMEROON

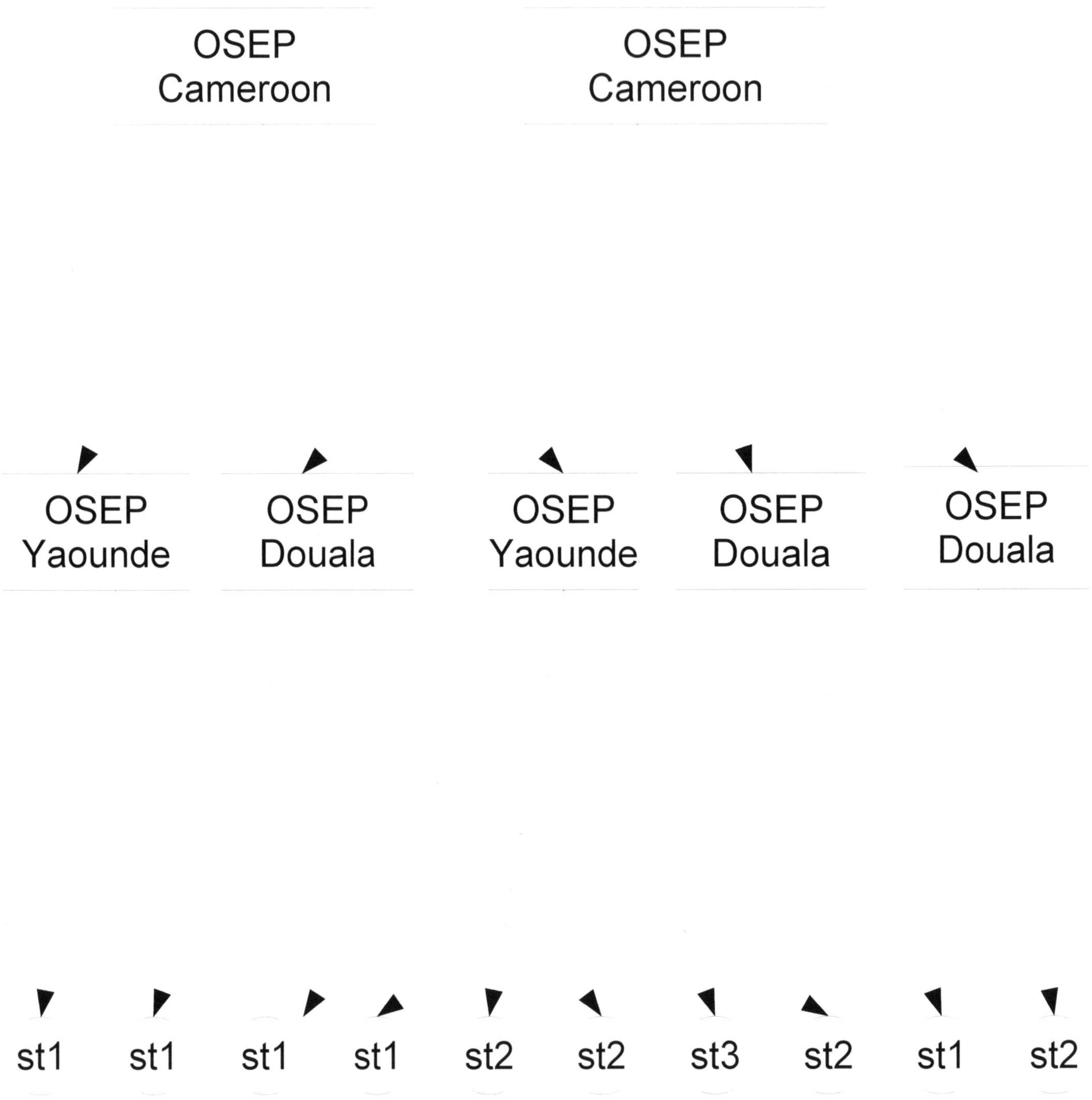

OSEP Cameroon

OSEP Cameroon

OSEP Yaounde

OSEP Douala

OSEP Yaounde

OSEP Douala

OSEP Douala

st1 st1 st1 st1 st2 st2 st3 st2 st1 st2

The OSEP central control in Cameroon Need the situational report of every OSEP office in the Yaounde and Douala

Msc.ing Feuzeu Yomsi
Raymond Jacquino

X1.C.ANALYSE OF OBAMA STIMULUS EDUCATION PLAN EFFECT IN 4 YEARS In CAMEROON

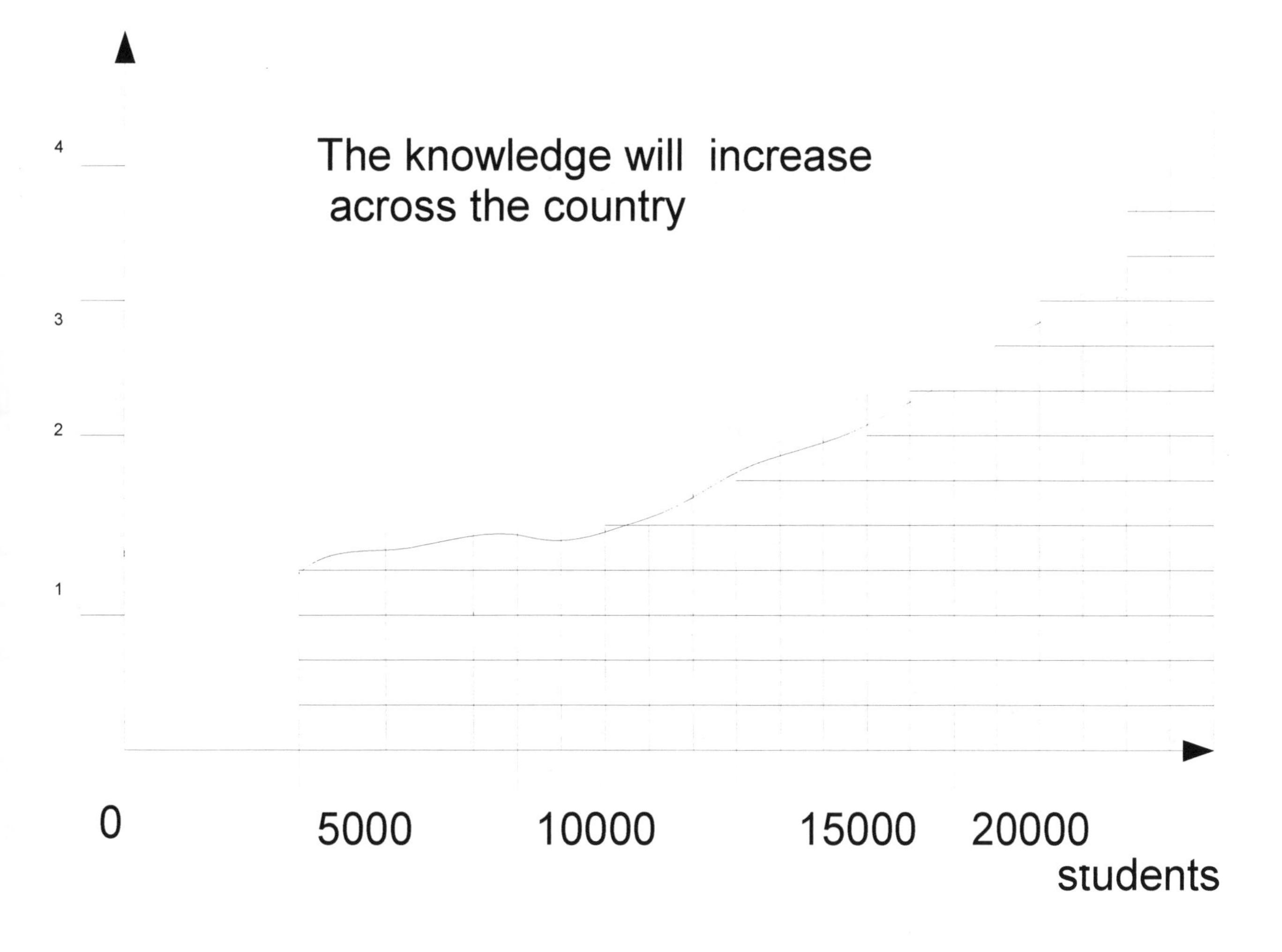

Msc.ing Feuzeu Yomsi
Raymond Jacquino

XI. TEST AND ANALYSE OF THE OSEP IN FRANCE

A. Entry Data

a) number of City =2 :PARIS and MARSEILLE

1. **FRANCE**
   city : PARIS
2. **FRANCE**
   City :MARSEILLE

b) number of OBAMA STIMULUS EDUCATION office in and FRANCE=5

OSEP/FRANCE/MARSEILLE/office1=id1=100a
OSEP/FRANCE/MARSEILLE/office2=id2=101a
OSEP/FRANCE/PARIS/office1=id3=200b
OSEP/FRANCE/PARIS/office2=id4=201b
OSEP/FRANCE/PARIS/office3=id5=202b

c)number of students who receive a consultation at OBAMA stimulus education plan office in Marseille and in Paris = 10

d)students receive appropriate consultation in :

1.mathematic in Marseille
2.Information technology in Paris

B. The central administration in FRANCE need the situational report of the OSEP across the country in this particular case in Marseille and Paris Eventually needed to know the average of satisfy students or not satisfy by the OSEP.

Our goal is to minimize the academy difficulties face by the students, maximize the success , increase the level of knowledge .

C. Graphical Interpretation

XI.TEST AND ANALYSE OF THE OSEP EFFECT IN FRANCE

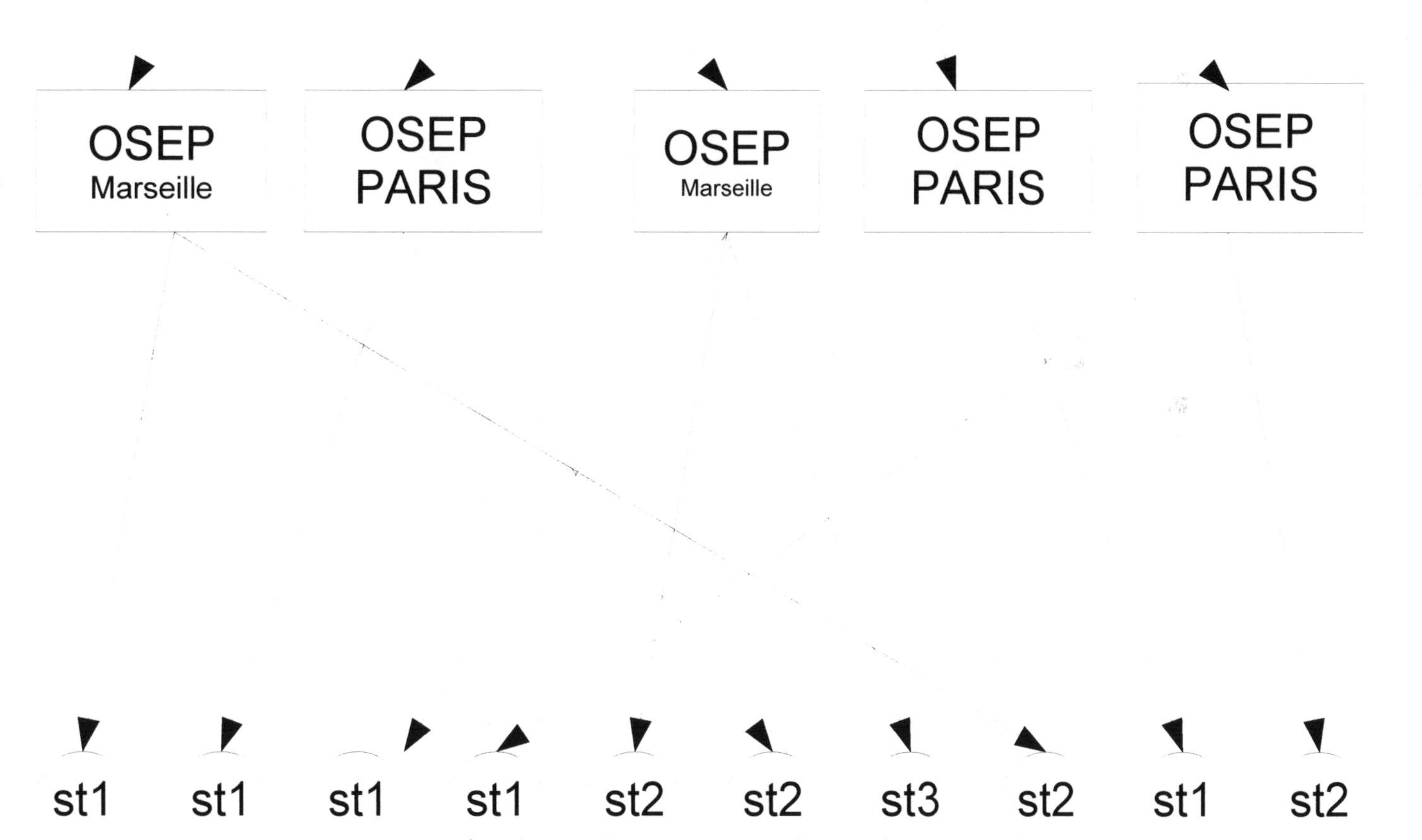

The OSEP central control in France Need the situational report of every OSEP office in Paris and Marseille

Msc.ing Feuzeu Yomsi
Raymond Jacquino

X1.C.ANALYSE OF OBAMA STIMULUS EDUCATION PLAN EFFECT IN 4 YEARS In FRANCE

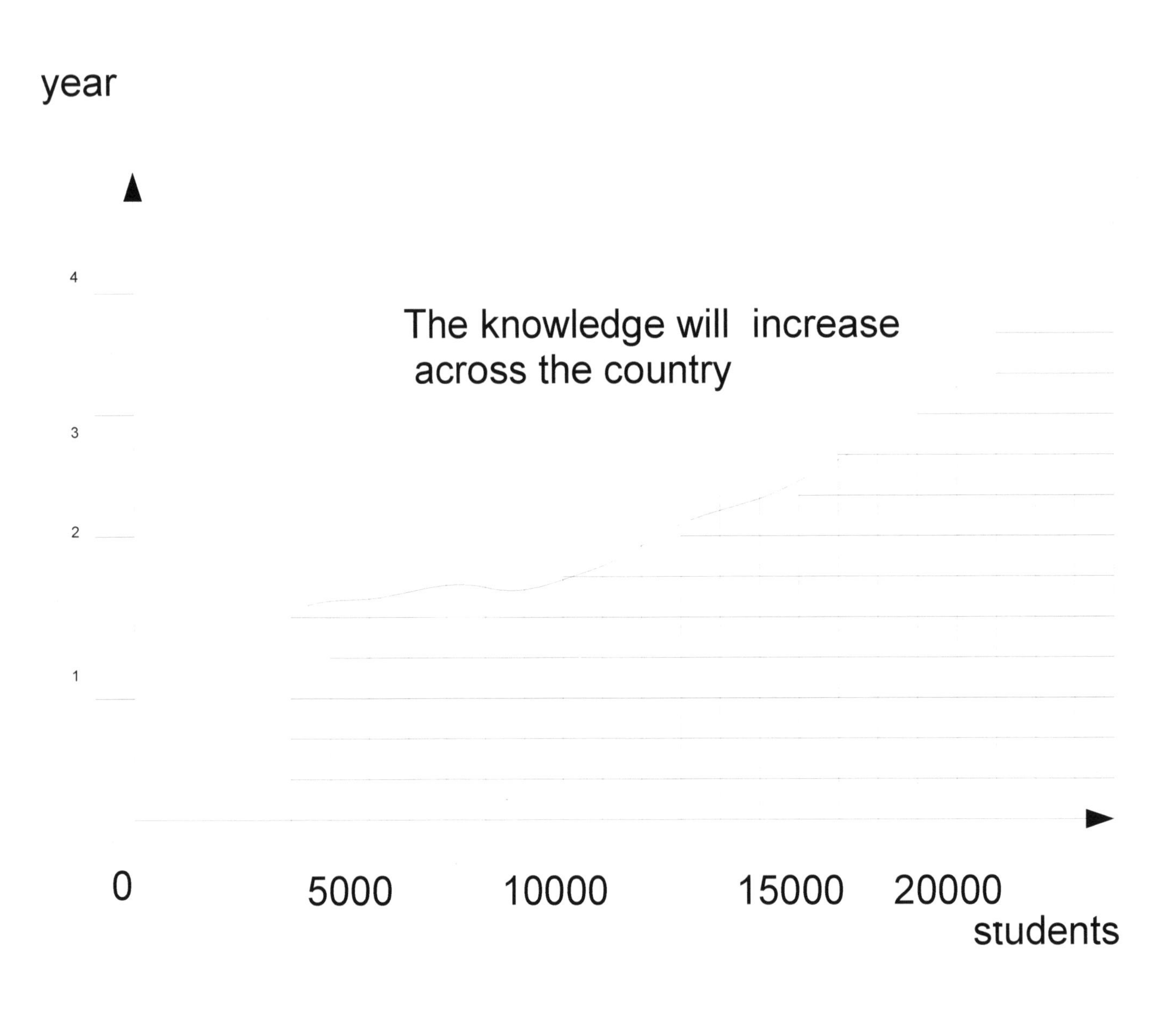

Msc.ing Feuzeu Yomsi
Raymond Jacquino

XI. TEST AND ANALYSE OF THE OSEP IN RUSSIA

A. Entry Data

a) number of City =2 : MOSCOW and ST PETERBURG

1. RUSSIA
   city : MOSCOW
2. RUSSIA
   City :ST PETERBURG

b) number of OBAMA STIMULUS EDUCATION office in Moscow and St Peterburg=5

OSEP/Russia/Moscow/office1=id1=100a
OSEP/Russia/Moscow/office2=id2=101a
OSEP/Russia/St Peterburg/office1=id3=200b
OSEP/Russia/St Peterburg/office2=id4=201b
OSEP/Russia/St Peterburg/office3=id5=202b

c)number of students who receive a consultation at OBAMA stimulus education plan office in Moscow and in St Peterburg= 10

d)students receive appropriate consultation in :

1.mathematic in Moscow
2.Information technology in St Peterburg

B. The central administration in Russia need the situational report of the OSEP across the country in this particular case in Moscow and St Peterburg. Eventually needed to know the average of satisfy students or not satisfy by the OSEP.

Our goal is to minimize the academy difficulties face by the students, maximize the success , increase the level of knowledge .

C. Graphical Interpretation

XI.TEST AND ANALYSE OF THE OSEP EFFECT In RUSSIA

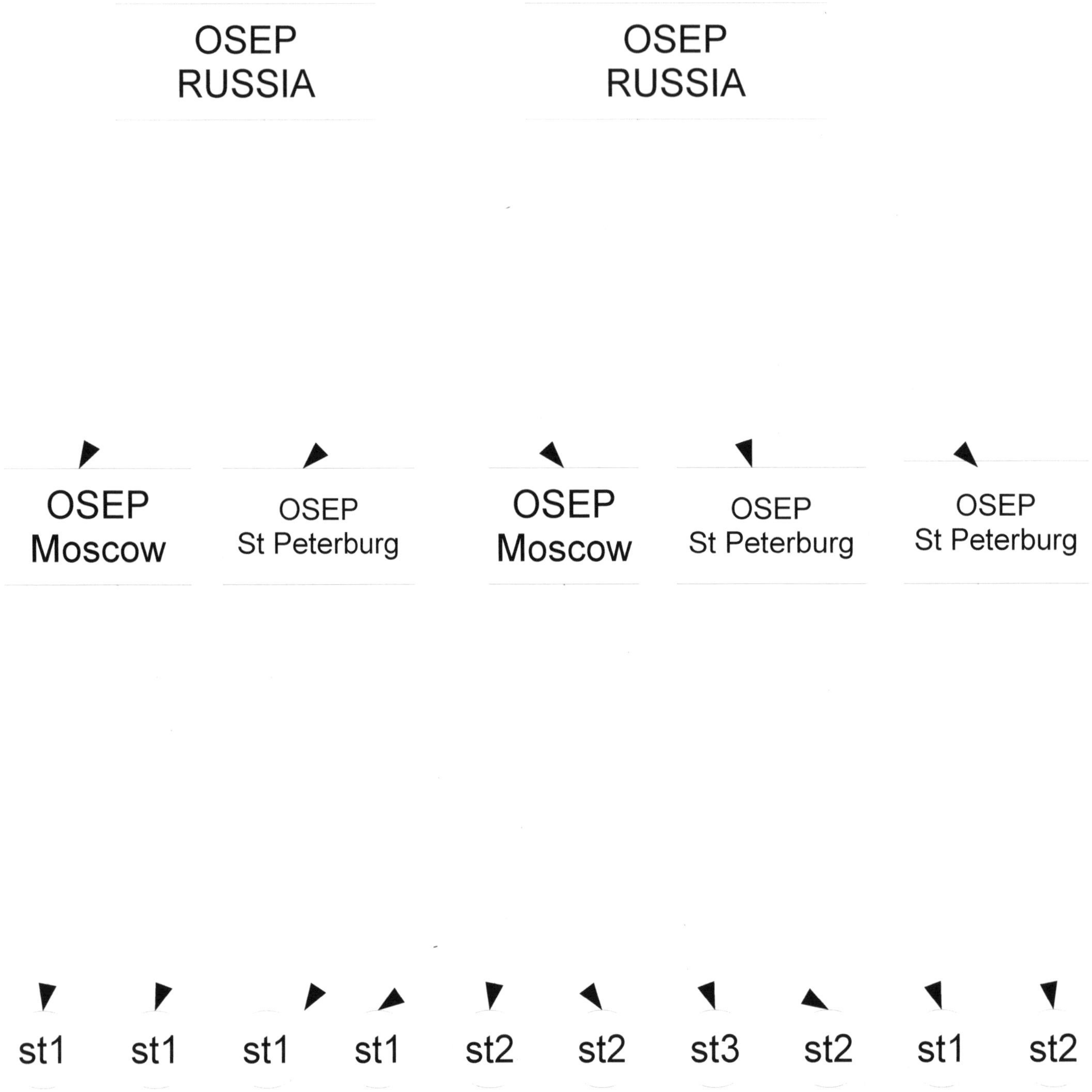

OSEP RUSSIA

OSEP RUSSIA

OSEP Moscow

OSEP St Peterburg

OSEP Moscow

OSEP St Peterburg

OSEP St Peterburg

st1 st1 st1 st1 st2 st2 st3 st2 st1 st2

The OSEP central control in Russia Need the situational report of every OSEP office in Moscow and StPeterburg

Msc.ing Feuzeu Yomsi
Raymond Jacquino

X1.C.ANALYSE OF OBAMA STIMULUS EDUCATION PLAN EFFECT IN 4 YEARS In RUSSIA

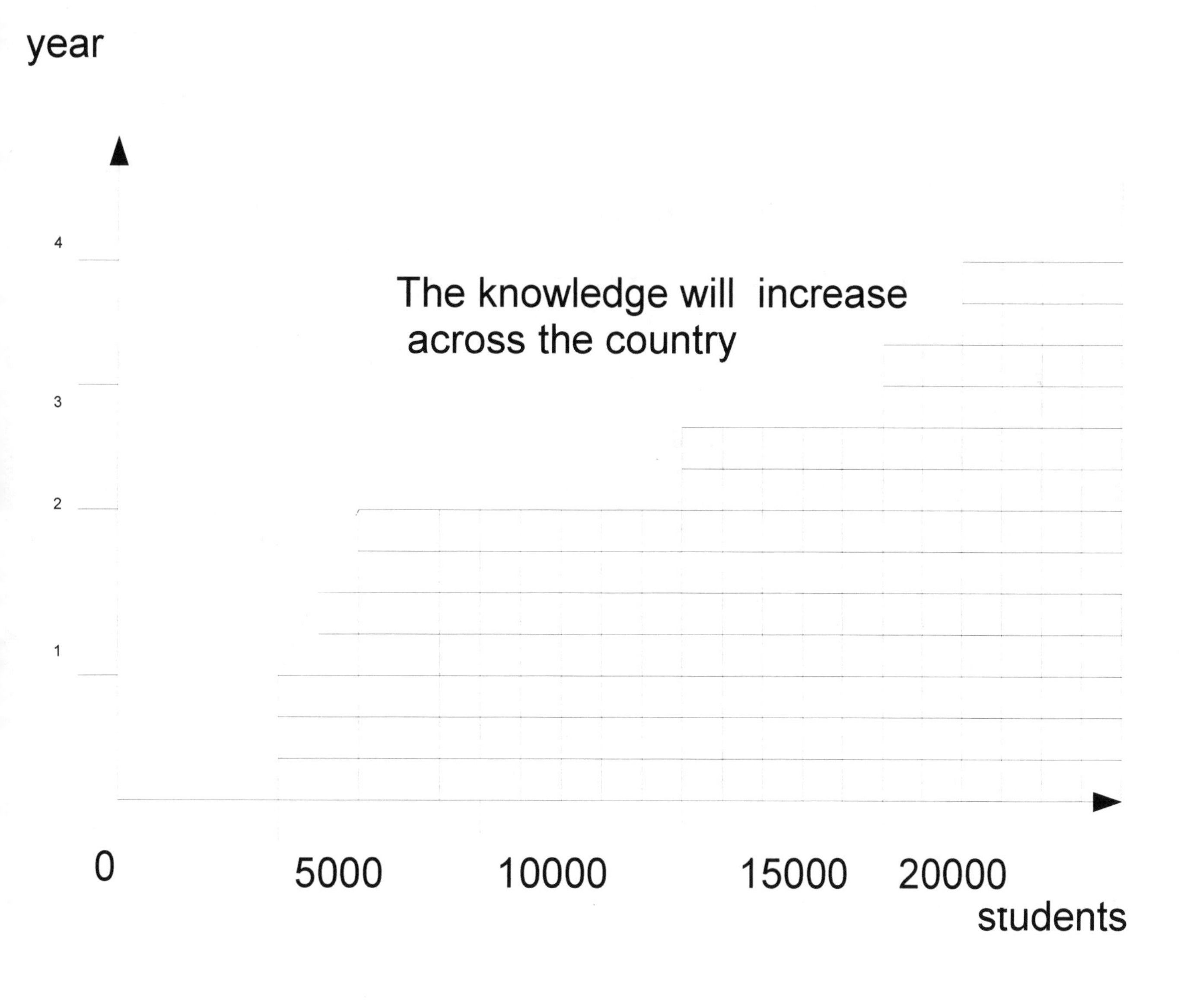

Msc.ing Feuzeu Yomsi
Raymond Jacquino

XI. TEST AND ANALYSE OF THE OSEP in INDONESIA

A. Entry Data

a) number of City =2 : JAKARTA and BALI

1. INDONESIA
   city : Jakarta
2. INDONESIA
   City :Bali

b) number of OBAMA STIMULUS EDUCATION office in and INDONESIA=5

OSEP/Indonesia/Jakarta/office1=id1=100a
OSEP/Indonesia/Jakarta/office2=id2=101a
OSEP/Indonesia/Bali/office1=id3=200b
OSEP/Indonesia/Bali/office2=id4=201b
OSEP/Indonesia/Bali/office3=id5=202b

c)number of students who receive a consultation at OBAMA stimulus education plan office in Jakarta and in Bali= 10

d)students receive appropriate consultation in :

1.mathematic in Jakarta
2.Information technology in Bali

B. The central administration in INDONESIA need the situational report of the OSEP across the country in this particular case in Jakarta and Bali.

Eventually needed to know the average of satisfy students or not satisfy by the OSEP.

Our goal is to minimize the academy difficulties face by the students, maximize the success , increase the level of knowledge .

C. Graphical Interpretation

XI.TEST AND ANALYSE OF THE OSEP EFFECT In INDONESIA

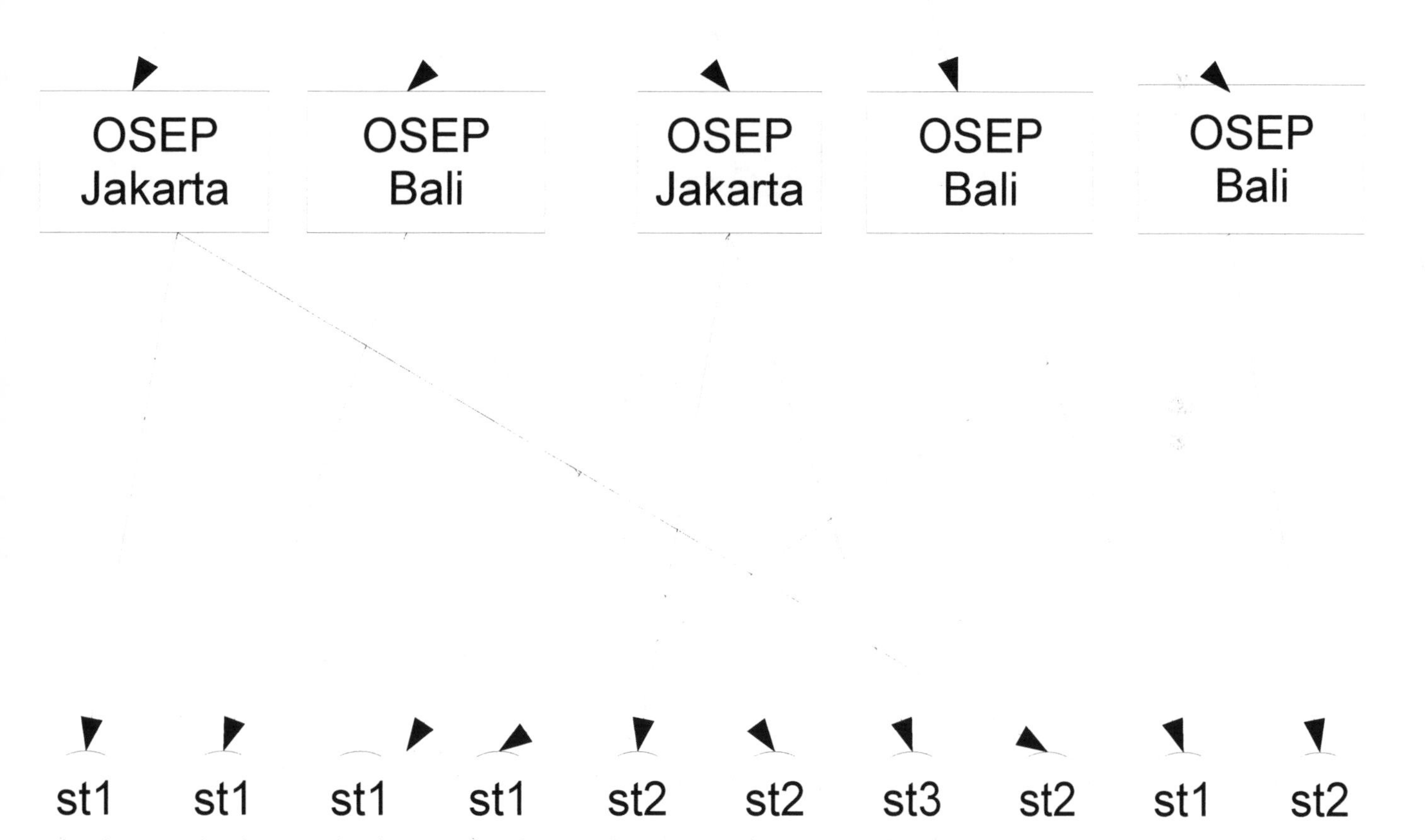

The OSEP central control in Indonesia Need the situational report of every OSEP office in Jakarta and Bali

Msc.ing Feuzeu Yomsi
Raymond Jacquino

X1.C.ANALYSE OF OBAMA STIMULUS EDUCATION PLAN EFFECT IN 4 YEARS In INDONESIA

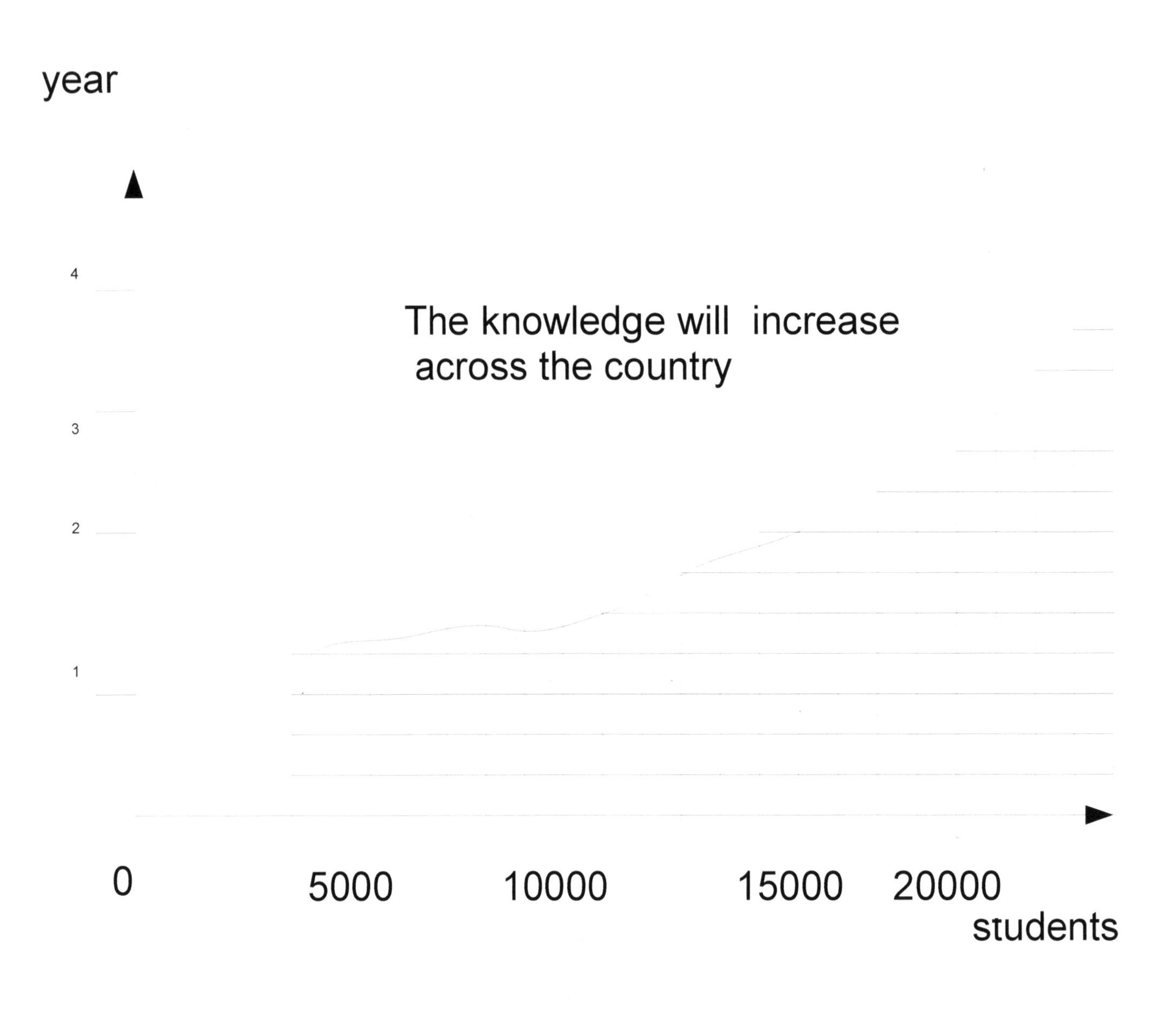

Msc.ing Feuzeu Yomsi
Raymond Jacquino

XI. TEST AND ANALYSE OF THE OSEP in INDIA

A. Entry Data

a) number of City =2 : DELHI and CALCUTTA

1. **INDIA**
   city : Delhi
2. **INDIA**
   City :Calcutta

b) number of OBAMA STIMULUS EDUCATION office in and INDIA=5

OSEP/India/Delhi/office1=id1=100a
OSEP/India/Delhi/office2=id2=101a
OSEP/India/Calcutta/office1=id3=200b
OSEP/India/Calcutta/office2=id4=201b
OSEP/India/Calcutta/office3=id5=202b

c)number of students who receive a consultation at OBAMA stimulus education plan office in Delhi and in Calcutta= 10

d)students receive appropriate consultation in :

1.mathematic in Delhi
2.Information technology in Calcutta

B. The central administration in INDIA need the situational report of the OSEP across the country in this particular case in Delhi and Calcutta.
Eventually needed to know the average of satisfy students or not satisfy by the OSEP.

Our goal is to minimize the academy difficulties face by the students, maximize the success , increase the level of knowledge .

C. Graphical Interpretation

XI.TEST AND ANALYSE OF THE OSEP EFFECT IN INDIA

OSEP
INDIA

OSEP
INDIA

OSEP
Calcutta

OSEP
DELHI

OSEP
Calcutta

OSEP
DELHI

OSEP
DELHI

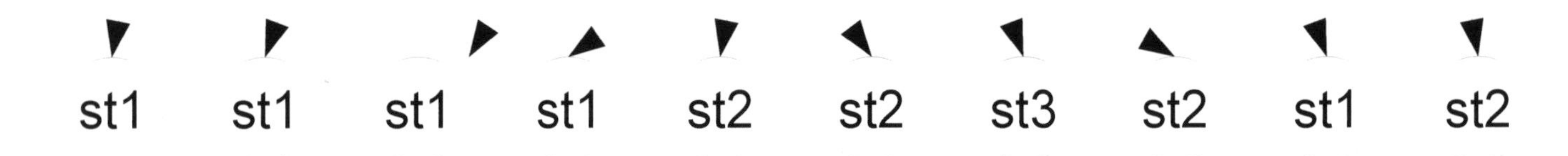

The OSEP central control in INDIA Need the situational report of every OSEP office in DELHI and CALCUTTA

Msc.ing Feuzeu Yomsi
Raymond Jacquino

X1.C.ANALYSE OF OBAMA STIMULUS EDUCATION PLAN EFFECT IN 4 YEARS In INDIA

year

4

3

2

1

The knowledge will increase
across the country

0 5000 10000 15000 20000
students

Msc.ing Feuzeu Yomsi
Raymond Jacquino

XI. TEST AND ANALYSE OF THE OSEP in CHINA

A. Entry Data

a) number of City =2 :Shanghai and Beijing

1. CHINA
   city : Shanghai
2. CHINA
   City :Beijing

b) number of OBAMA STIMULUS EDUCATION office in and CHINA=5

OSEP/China/Shanghai/office1=id1=100a
OSEP/China/Shanghai/office2=id2=101a
OSEP/China/Beijing/office1=id3=200b
OSEP/China/Beijing/office2=id4=201b
OSEP/China/Beijing/office3=id5=202b

c)number of students who receive a consultation at OBAMA stimulus education plan office in Shanghai and in Beijing= 10

d)students receive appropriate consultation in :

1.mathematic in Shanghai
2.Information technology in Beijing

B. The central administration in CHINA need the situational report of the OSEP across the country in this particular case in Shanghai and Beiging.

Eventually needed to know the average of satisfy students or not satisfy by the OSEP.

Our goal is to minimize the academy difficulties face by the students, maximize the success , increase the level of knowledge .

C. Graphical Interpretation

XI.TEST AND ANALYSE OF THE OSEP EFFECT In CHINA

OSEP
CHINA

OSEP
CHINA

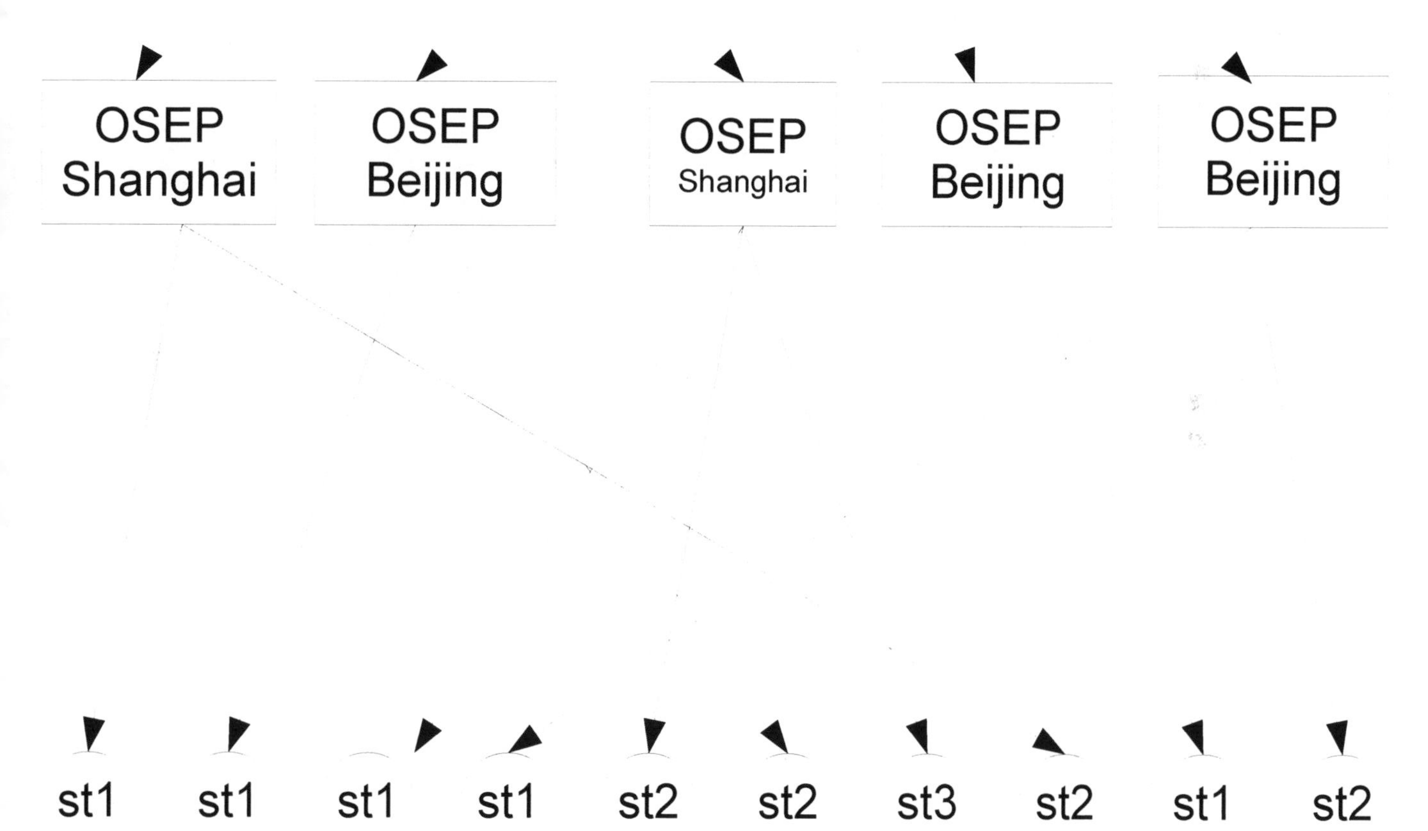

The OSEP central control in Indonesia Need the situational report of every OSEP office in Shanghai and Beijing

Msc.ing Feuzeu Yomsi
Raymond Jacquino

X1.C.ANALYSE OF OBAMA STIMULUS EDUCATION PLAN EFFECT IN 4 YEARS In CHINA

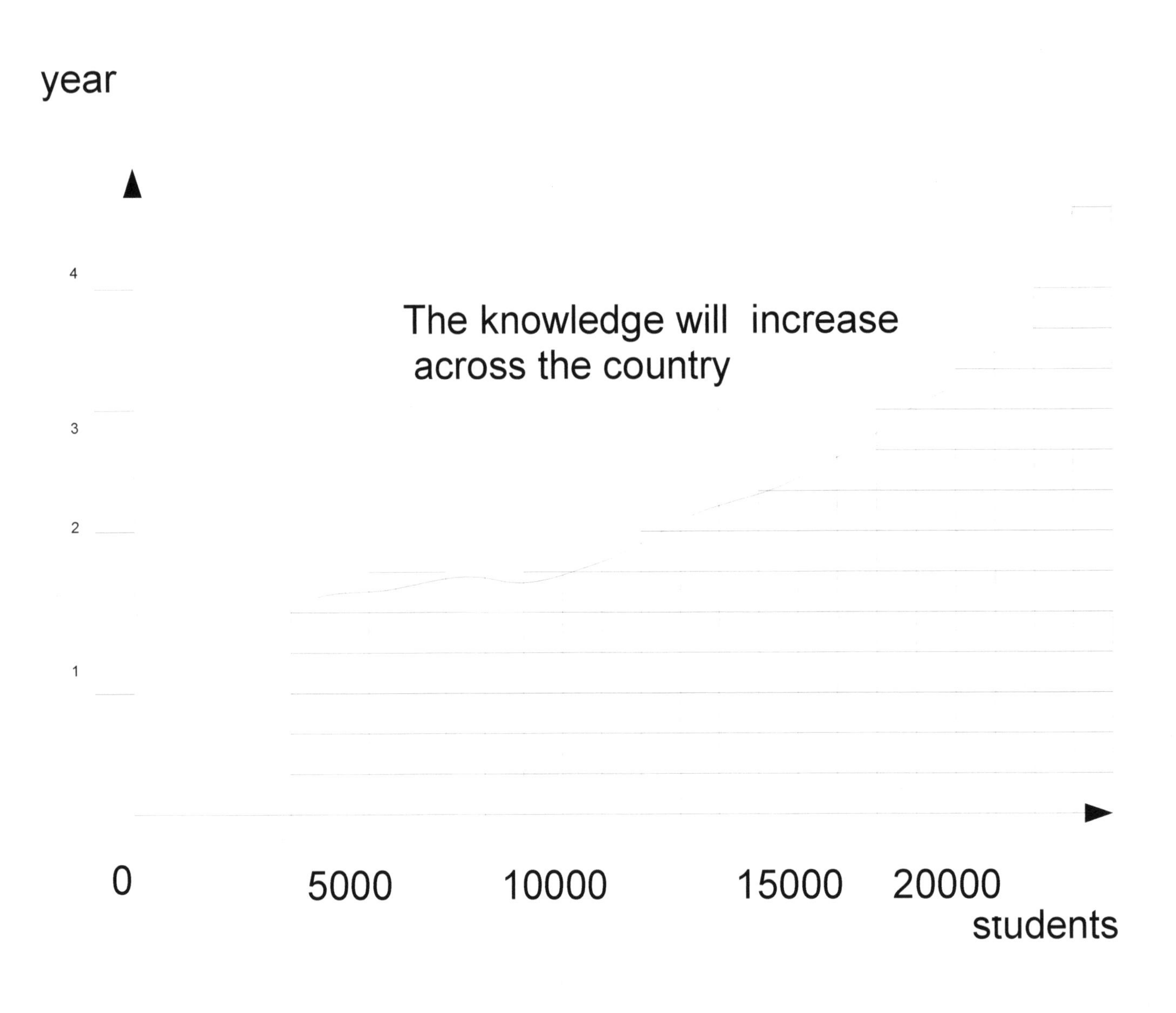

Msc.ing Feuzeu Yomsi
Raymond Jacquino

XI. TEST AND ANALYSE OF THE OSEP in JAPAN

A. Entry Data

a) number of City =2 :TOKYO and OSAKA

1. JAPAN
   city : Tokyo
2. JAPAN
   City :Osaka

b) number of OBAMA STIMULUS EDUCATION office in and Tokyo=5

OSEP/Japan/Tokyo/office1=id1=100a
OSEP/Japan/Tokyo/office2=id2=101a
OSEP/Japan/Osaka/office1=id3=200b
OSEP/Japan/Osaka/office2=id4=201b
OSEP/Japan/Osaka/office3=id5=202b

c)number of students who receive a consultation at OBAMA stimulus education plan office in Tokyo and in Osaka= 10

d)students receive appropriate consultation in :

1.mathematic in Tokyo
2.Information technology in Osaka

B. The central administration in JAPAN need the situational report of the OSEP across the country in this particular case in Tokyo and Osaka
Eventually needed to know the average of satisfy students or not satisfy by the OSEP.

Our goal is to minimize the academy difficulties face by the students, maximize the success , increase the level of knowledge .

C. Graphical Interpretation

XI.TEST AND ANALYSE OF THE OSEP EFFECT In JAPAN

OSEP JAPAN

OSEP JAPAN

OSEP Osaka

OSEP TOKYO

OSEP Osaka

OSEP Osaka

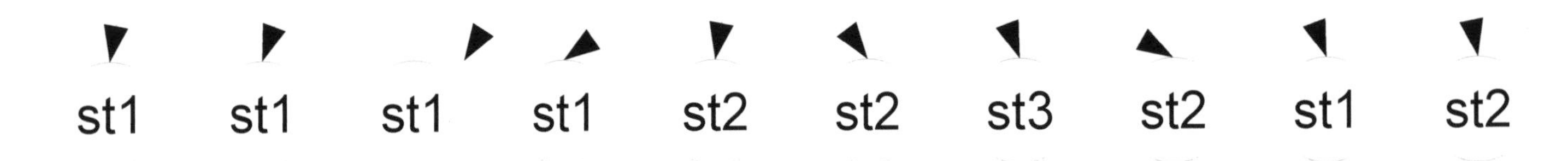

The OSEP central control in Indonesia Need the situational report of every OSEP office in Tokyo and Osaka

Msc.ing Feuzeu Yomsi
Raymond Jacquino

X1.C.ANALYSE OF OBAMA STIMULUS EDUCATION PLAN EFFECT IN 4 YEARS In 3APAN

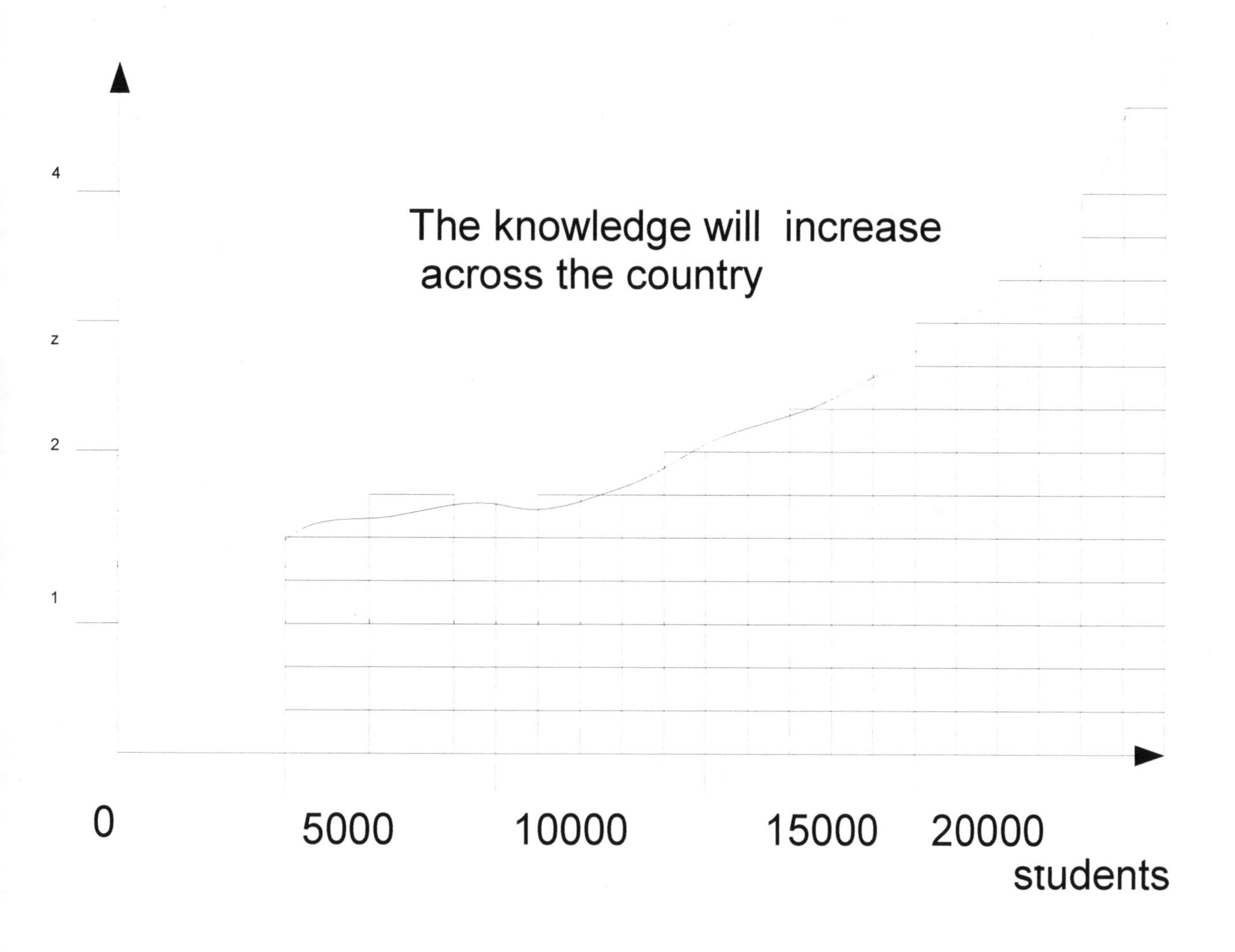

Msc.ing Feumeu YoJ si
RayJ ond 3acquino

XI. TEST AND ANALYSE OF THE OSEP in ENGLAND

A. Entry Data

a) number of City =2 :LONDON and LIVERPOOL

1. ENGLAND
   City : London
2. ENGLAND
   City :Liverpool

b) number of OBAMA STIMULUS EDUCATION office in London and Liverpool =5

OSEP/England/London/office1=id1=100a
OSEP/England/London/office2=id2=101a
OSEP/England/Liverpool/office1=id3=200b
OSEP/England/Liverpool/office2=id4=201b
OSEP/England/Liverpool/office3=id5=202b

c)number of students who receive a consultation at OBAMA stimulus education plan office in London and in Liverpool= 10

d)students receive appropriate consultation in :
1.mathematic in London
2.Information technology in Liverpool

B. The central administration in England need the situational report of the OSEP across the country in this particular case in London and Liverpool Eventually needed to know the average of satisfy students or not satisfy by the OSEP.

Our goal is to minimize the academy difficulties face by the students, maximize the success , increase the level of knowledge .

C. Graphical Interpretation

XI.TEST AND ANALYSE OF THE OSEP EFFECT In ENGLAND

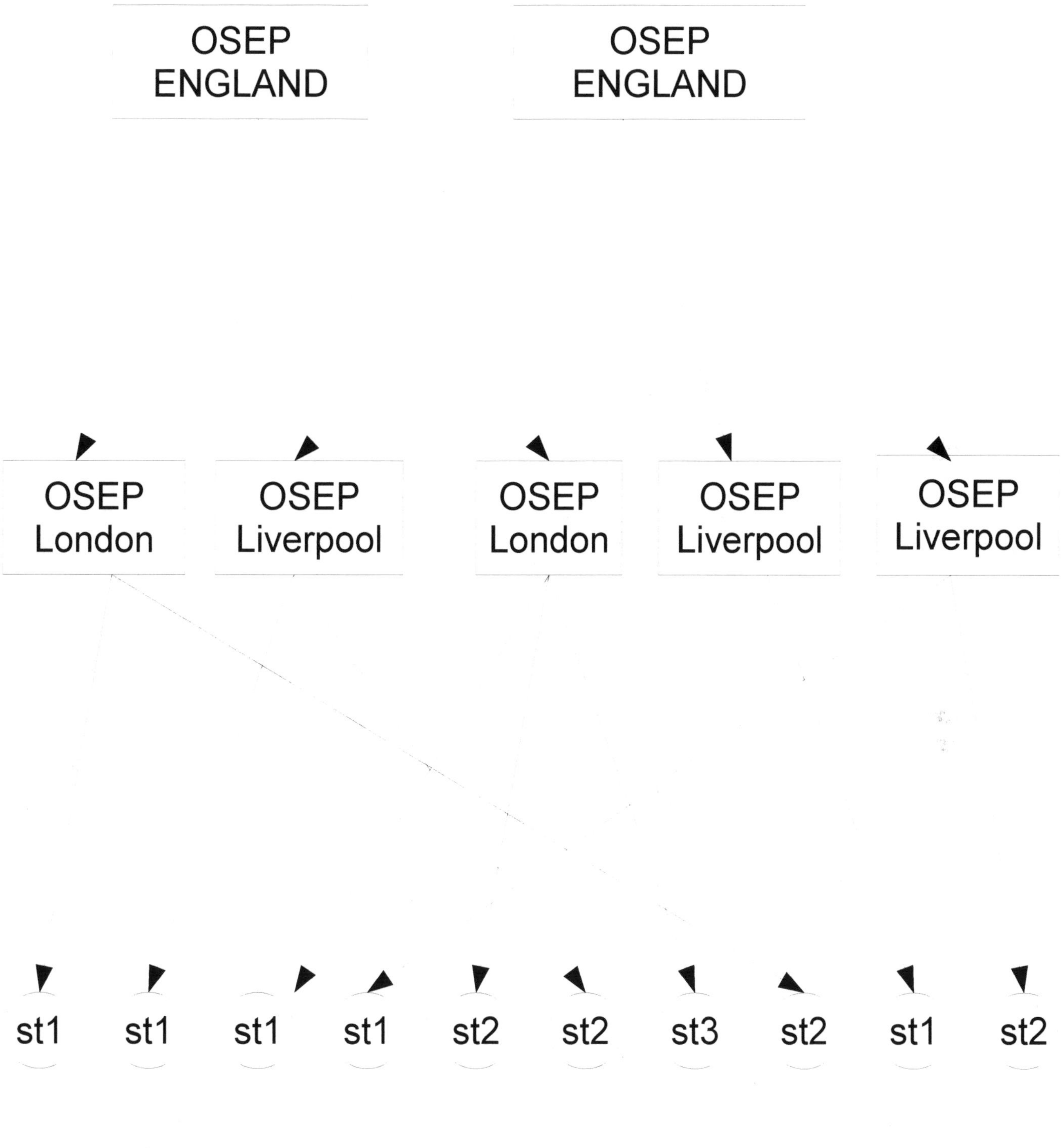

OSEP
ENGLAND

OSEP
ENGLAND

OSEP
London

OSEP
Liverpool

OSEP
London

OSEP
Liverpool

OSEP
Liverpool

st1 st1 st1 st1 st2 st2 st3 st2 st1 st2

The OSEP central control in ENGLAND Need the situational report of every OSEP office in London and Liverpool

Msc.ing Feuzeu Yomsi
Raymond Jacquino

X1.C.ANALYSE OF OBAMA STIMULUS EDUCATION PLAN EFFECT IN 4 YEARS In ENGLAND

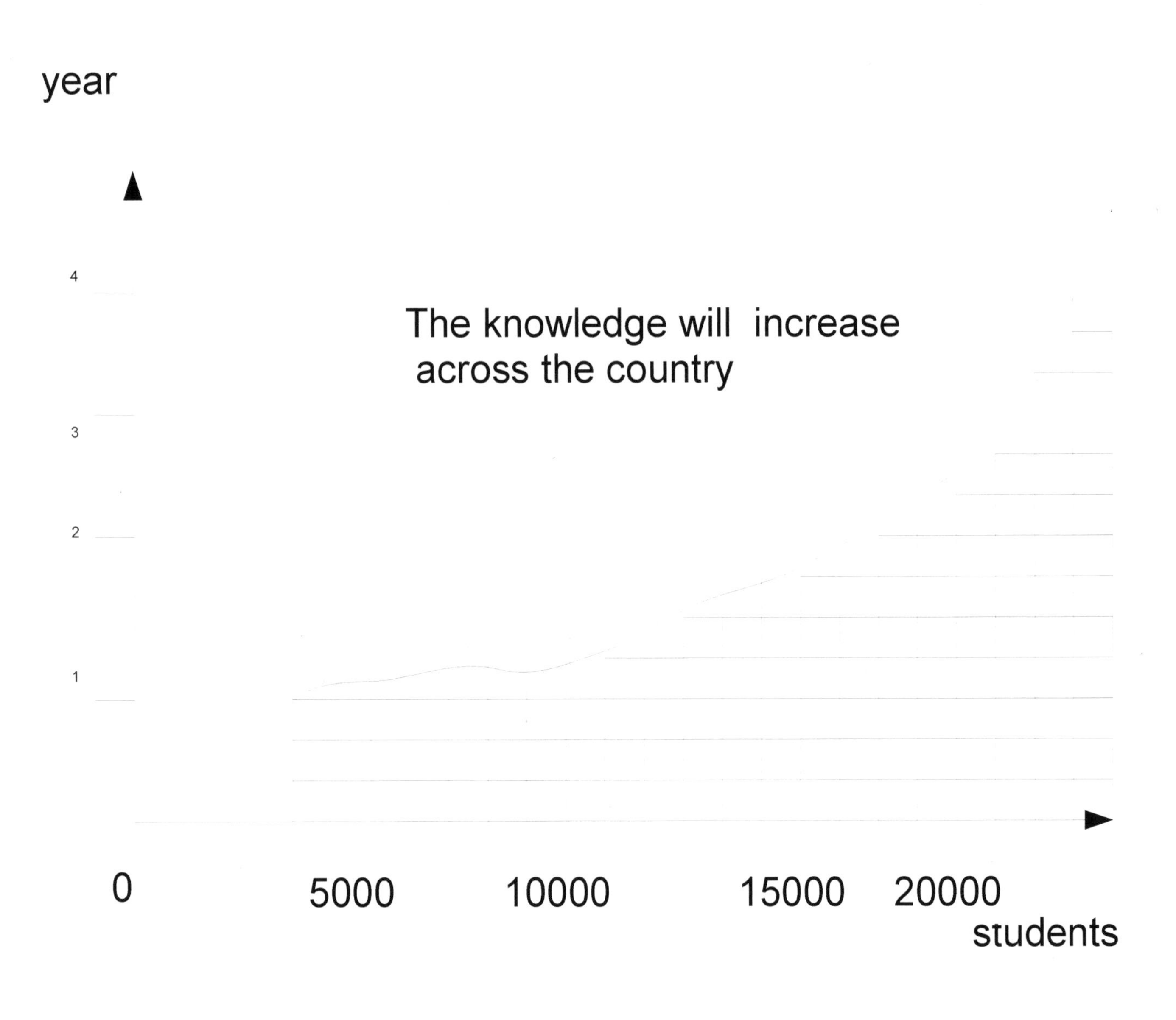

Msc.ing Feuzeu Yomsi
Raymond Jacquino

XI. TEST AND ANALYSE OF THE OSEP in BRASIL

A. Entry Data

a) number of City =2 :RIO DE JANEIRO and SAN PAULO

1. BRASIL
   City : RIO DE JANEIRO
2. BRASIL
   City :SAN PAULO

b) number of OBAMA STIMULUS EDUCATION office in Rio de Janeiro and San Paulo =5

OSEP/BRASIL/RIO DE JANEIRO/office1=id1=100a
OSEP/BRASIL/RIO DE JANEIRO/office2=id2=101a
OSEP/BRASIL/San Paulo/office1=id3=200b
OSEP/BRASIL/San Paulo/office2=id4=201b
OSEP/BRASIL/San Paulo/office3=id5=202b

c)number of students who receive a consultation at OBAMA stimulus education plan office in Rio de Janeiro and in San Paulo= 10

d)students receive appropriate consultation in :

1.mathematic in Rio de Janeiro
2.Information technology in San Paulo

B. The central administration in BRASIL need the situational report of the OSEP across the country in this particular case in Rio de Janeiro and San Paulo Eventually needed to know the average of satisfy students or not satisfy by the OSEP.

Our goal is to minimize the academy difficulties face by the students, maximize the success , increase the level of knowledge .

C. Graphical Interpretation

XI.TEST AND ANALYSE OF THE OSEP EFFECT In BRASIL

OSEP
BRASIL

OSEP
BRASIL

OSEP
San Paulo

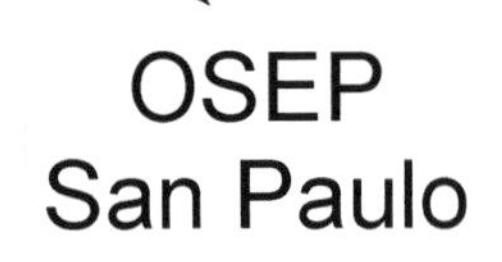

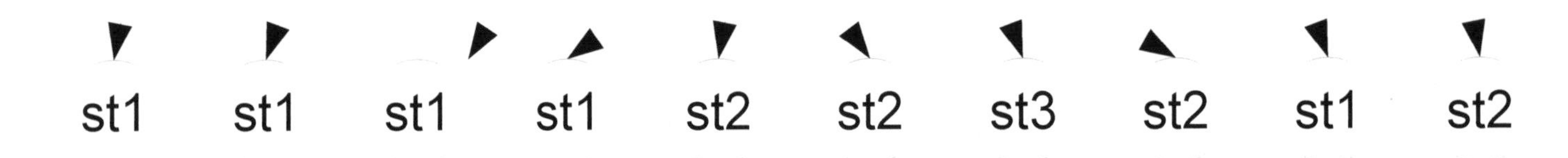

The OSEP central control in BRASIL Need the situational report of every OSEP office in Rio de Janeiro and San Paulo

Msc.ing Feuzeu Yomsi
Raymond Jacquino

X1.C.ANALYSE OF OBAMA STIMULUS EDUCATION PLAN EFFECT IN 4 YEARS In BRASIL

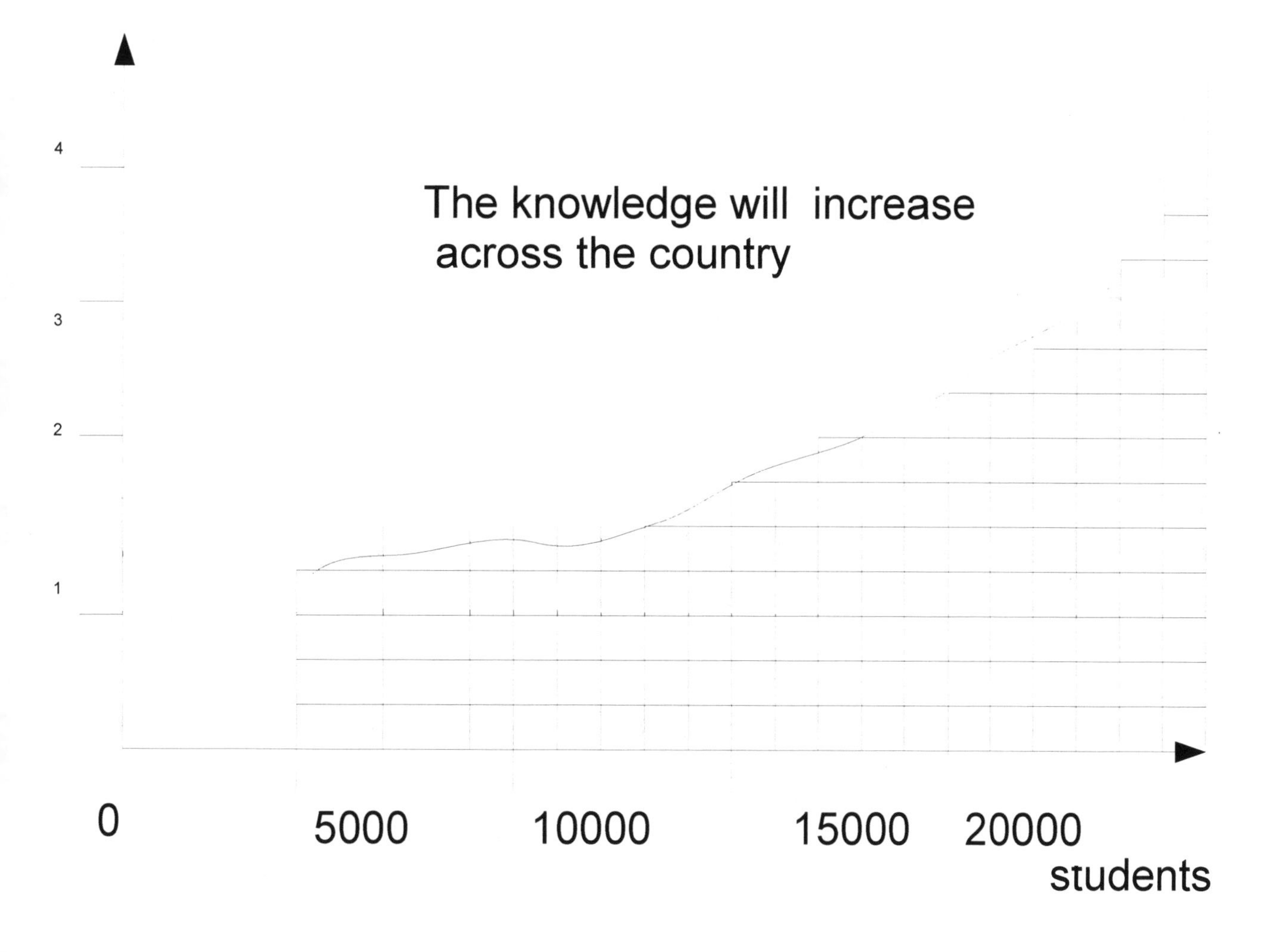

Msc.ing Feuzeu Yomsi
Raymond Jacquino

XI. TEST AND ANALYSE OF THE OSEP in ABRENTINA

A. Entry Data

a) number of City =2 :J d ENOS AIBES anU J ABILOCHE

XII. 1. ABRENTINA

City : J d ENOS AIBES

2. ABRENTINA

City :J ABILOCHE

b) number of OJ AMA STIMd Ld S EDd CATION office in Bio Ue l aneiro anU San Pau5o =/

OSEP0ABRENTINA0J d ENOS AIBES0office1=iU1=133a

OSEP0ABRENTINA0J d ENOS AIBES0office2=iU2=131a

OSEP0ABRENTINA0J ABILOCHE0office1=iU4=233b

OSEP0ABRENTINA0J ABILOCHE0office2=iUs =231b

OSEP0ABRENTINA0J ABILOCHE0office4=iU/ =232b

c)number of wtuUentwh vo receipe a conwu5tation at OJ AMA wtimu5uweUucation g5an office in J d ENOS AIBES anU in J ABILOCHE= 13

U)wtuUentwreceipe aggrogriate conwu5tation in :

1.matvematic in Bio J d ENOS AIBES

2.Information tecvno5oky in J ABILOCHE

J . Tve centra5 aUminiwtration in ABRENTINA neeU tve wituationa5regort of tve OSEP acroww tve country in tviwgarticu5ar cawe in Bio Ue l aneiro anU San Pau5o Epentua55y neeUeU to z noh tve aperake of watiwfy wtuUentwor not watiwfy by tve OSEP.

Our koa5iwto minimi, e tve acaUemy Uifficu5tiewface by tve wtuUentwx maGmi, e tve wuccewwxincreawe tve 5epe5of z noh 5eUke .

C. Rragvica5Intergretation

XI.TEST AND ANALYSE OF THE OSEP EFFECT In ARGENTINA

OSEP
ARGENTINA

OSEP
ARGENTINA

OSEP
Bariloche

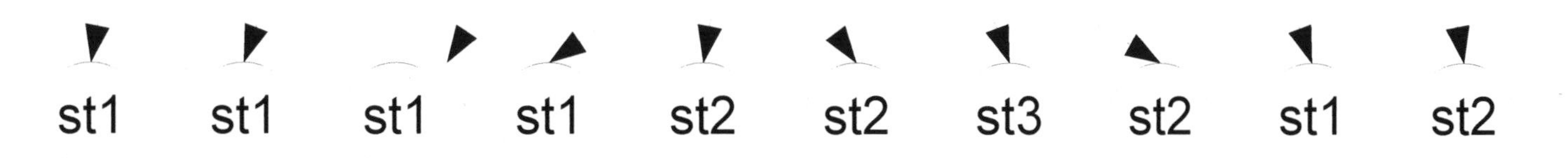

The OSEP central control in ARGENTINA Need the situational report of every OSEP office in Buenos aires and Bariloche

Msc.ing Feuzeu Yomsi
Raymond Jacquino

X1.C.ANALYSE OF OBAMA STIMULUS EDUCATION PLAN EFFECT IN 4 YEARS In ARGENTINA

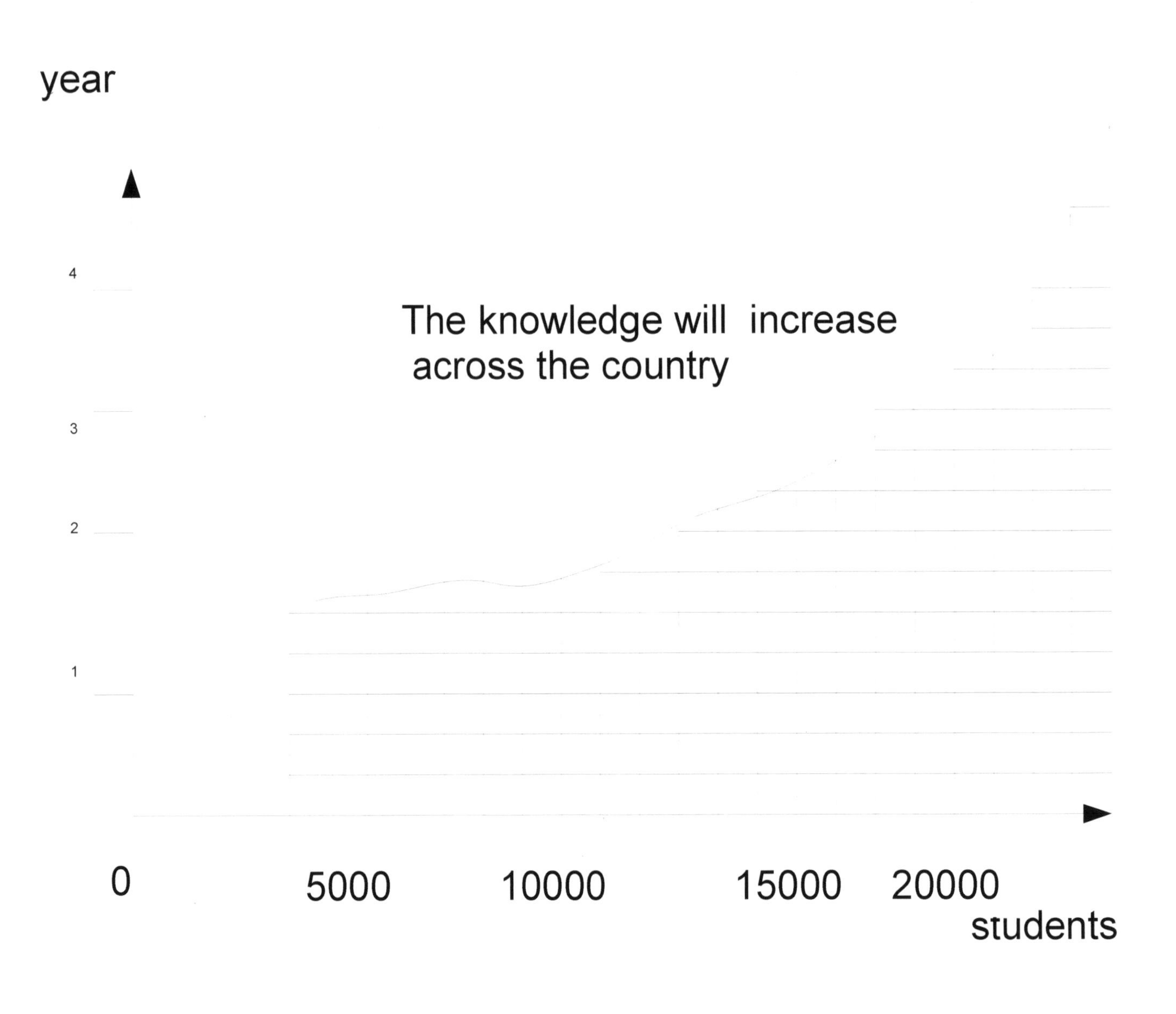

Msc.ing Feuzeu Yomsi
Raymond Jacquino

XI. TEST AND ANALYSE OF THE OSEP in ISRAEL

A. Entry Data

a) number of City =2 TEL AVIV AND JERUSALEM

XII. 1. ISRAEL

City : TEL AVIV

2. ISRAEL

City :JERUSALEM

b) number of OBAMA STIMULUS EDUCATION office in TEL AVIV AND Jerusalem =5

OSEP/ISRAEL/TEL AVIV/office1=id1=100a
OSEP/ISRAEL/TEL AVIV/office2=id2=101a
OSEP/ISRAEL/JERUSALEM/office1=id3=200b
OSEP/ISRAEL/JERUSALEM/office2=id4=201b
OSEP/ISRAEL/JERUSALEM/office3=id5=202b

c)number of students who receive a consultation at OBAMA stimulus education plan office in TEL AVIV and in JERUSALEM = 10

d)students receive appropriate consultation in :

1.mathematic in TEL AVIV
2.Information technology in JERUSALEM

B. The central administration in ISRAEL need the situational report of the OSEP across the country in this particular case in TEL AVIV and JERUSALEM Eventually needed to know the average of satisfy students or not satisfy by the OSEP.

Our goal is to minimize the academy difficulties face by the students, maximize the success , increase the level of knowledge .

C. Graphical Interpretation

XI.TEST AND ANALYSE OF THE OSEP EFFECT In SOUTH AFRICA

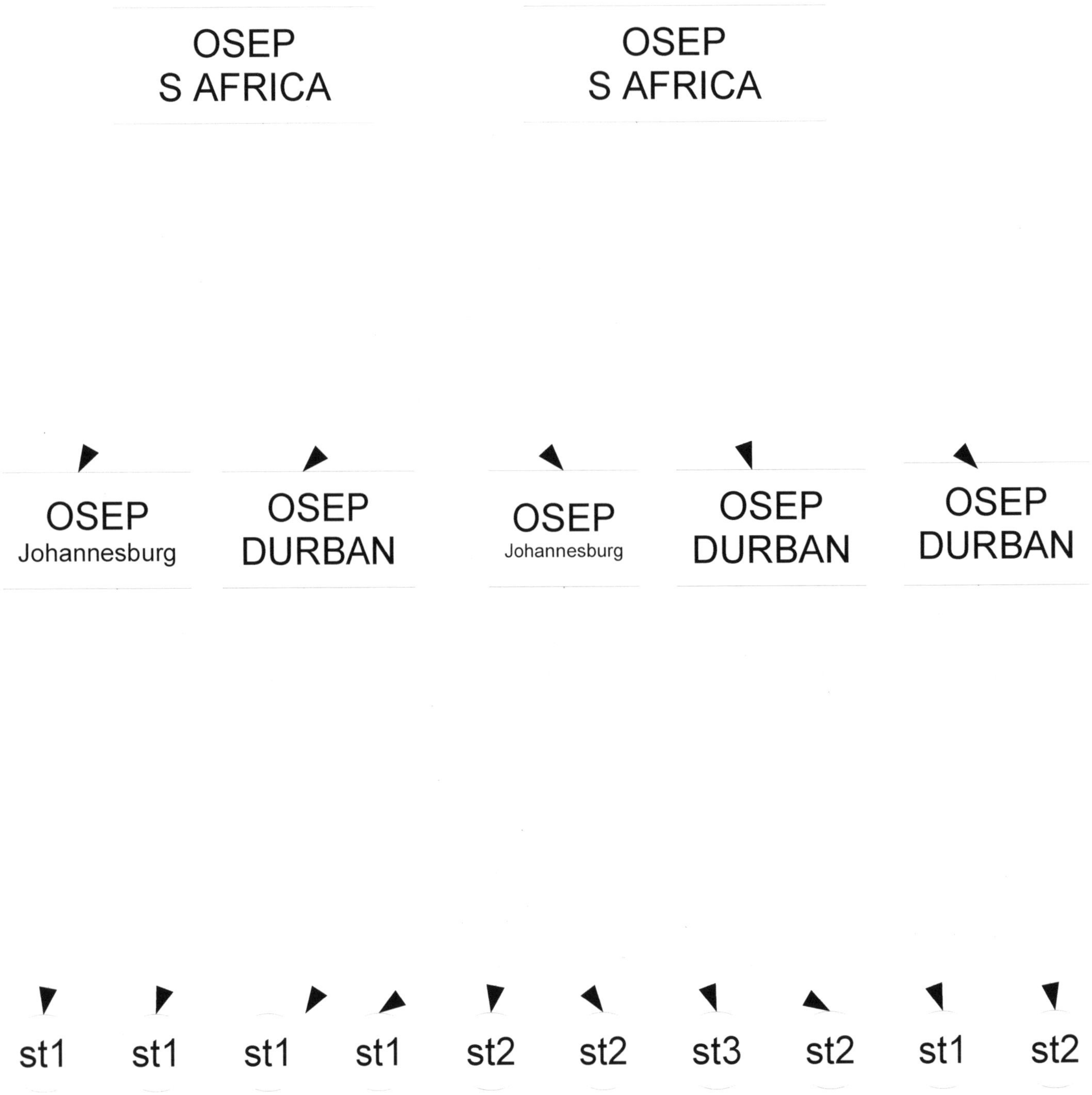

OSEP
S AFRICA

OSEP
S AFRICA

OSEP
Johannesburg

OSEP
DURBAN

OSEP
Johannesburg

OSEP
DURBAN

OSEP
DURBAN

st1 st1 st1 st1 st2 st2 st3 st2 st1 st2

The OSEP central control in South Africa Need the situational report of every OSEP office in Johannesburg and Durban

Msc.ing Feuzeu Yomsi
Raymond Jacquino

X1.C.ANALYSE OF OBAMA STIMULUS EDUCATION PLAN EFFECT IN 4 YEARS In SOUTH AFRICA

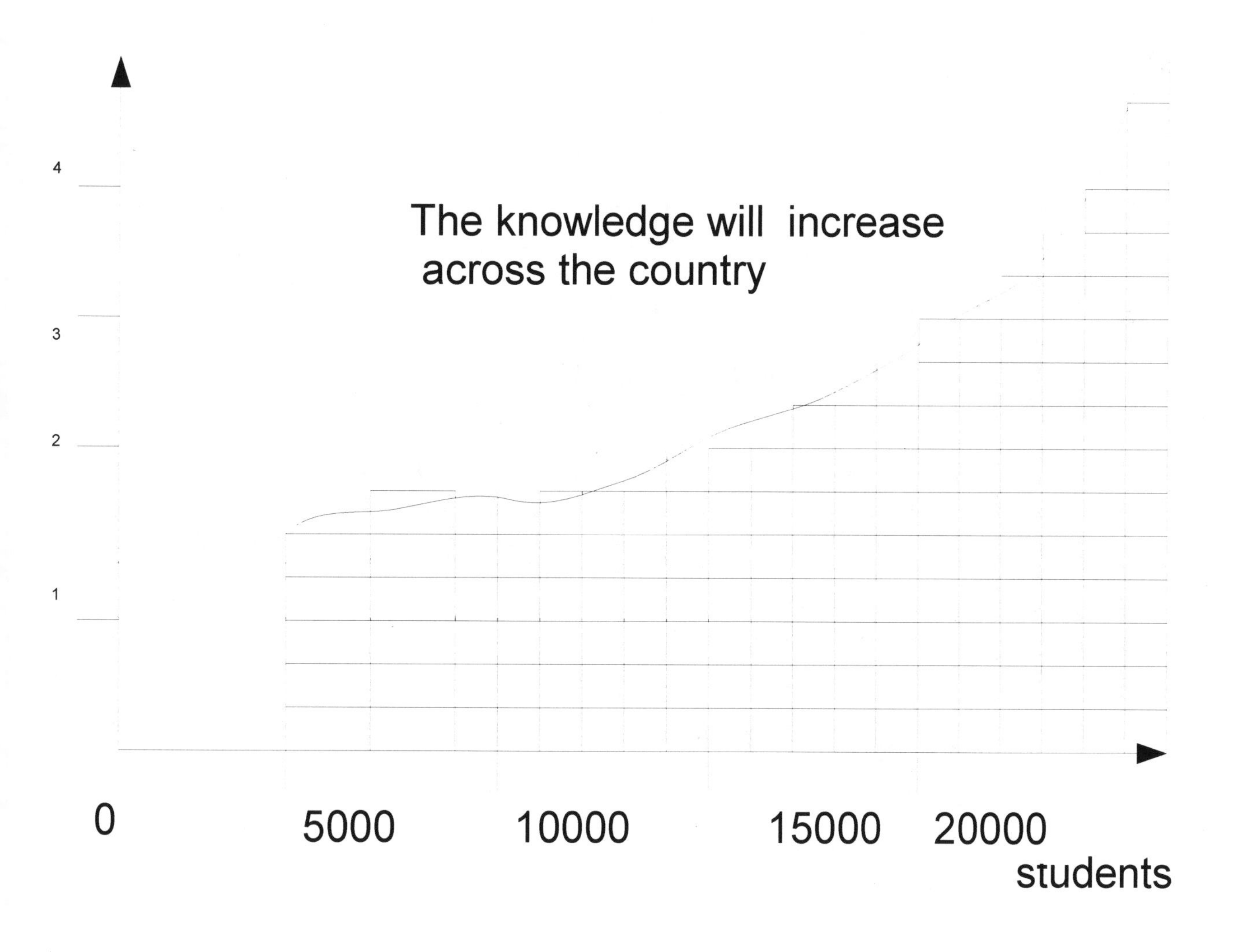

Msc.ing Feuzeu Yomsi
Raymond Jacquino

XI. TEST AND ANALYSE OF THE OSEP in ISRAEL

A. Entry Data

a) number of City =2 TEL AVIV AND JERUSALEM

XII.
1. **ISRAEL**
   City : TEL AVIV
2. **ISRAEL**
   City :JERUSALEM

b) number of OBAMA STIMULUS EDUCATION office in TEL AVIV AND Jerusalem =5

OSEP/ISRAEL/TEL AVIV/office1=id1=100a
OSEP/ISRAEL/TEL AVIV/office2=id2=101a
OSEP/ISRAEL/JERUSALEM/office1=id3=200b
OSEP/ISRAEL/JERUSALEM/office2=id4=201b
OSEP/ISRAEL/JERUSALEM/office3=id5=202b

c)number of students who receive a consultation at OBAMA stimulus education plan office in TEL AVIV and in JERUSALEM = 10

d)students receive appropriate consultation in :

1.mathematic in TEL AVIV
2.Information technology in JERUSALEM

B. The central administration in ISRAEL need the situational report of the OSEP across the country in this particular case in TEL AVIV and JERUSALEM Eventually needed to know the average of satisfy students or not satisfy by the OSEP.

Our goal is to minimize the academy difficulties face by the students, maximize the success , increase the level of knowledge .

C. Graphical Interpretation

XI.TEST AND ANALYSE OF THE OSEP EFFECT In ISRAEL

OSEP ISRAEL

OSEP ISRAEL

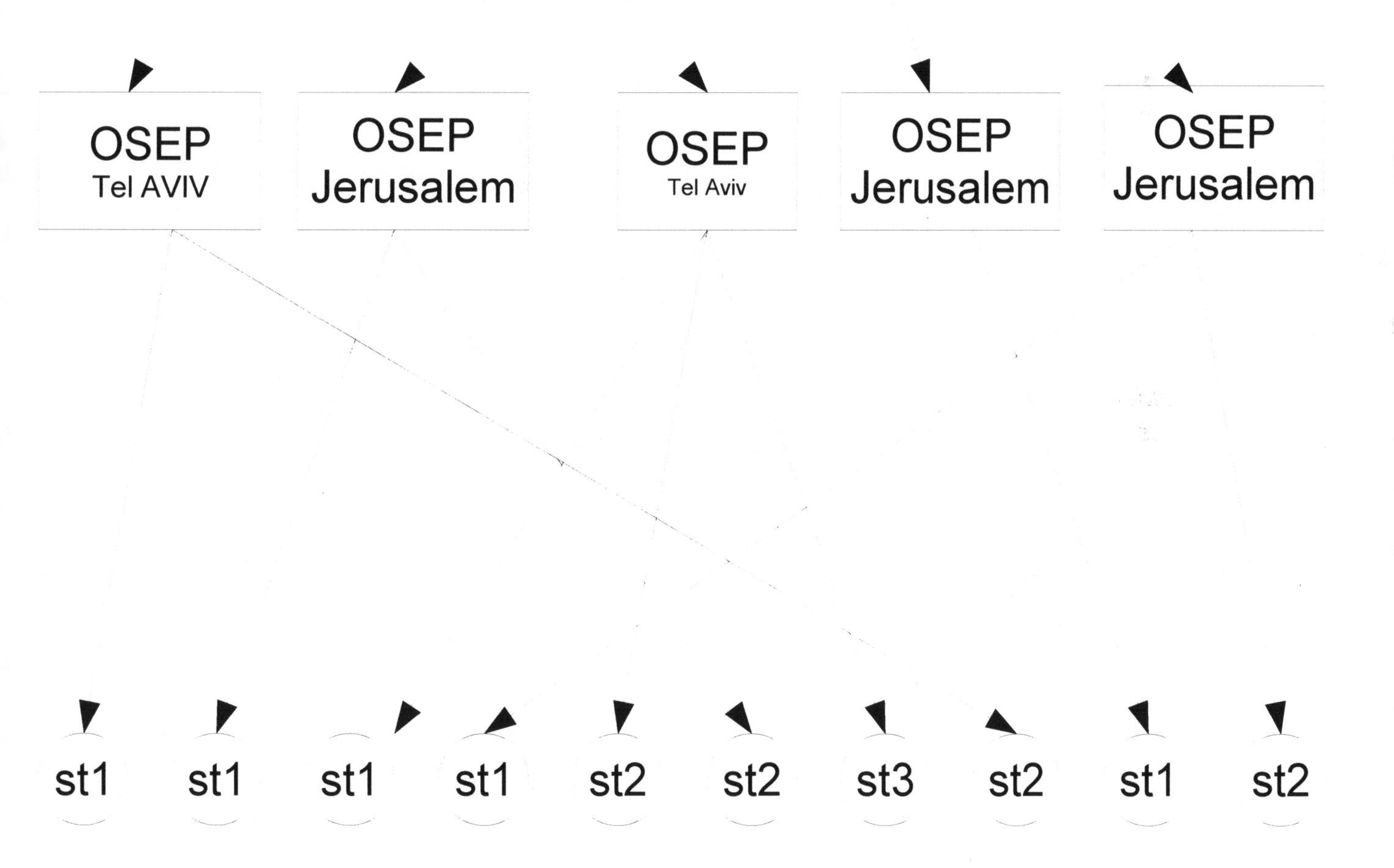

The OSEP central control in Israel Need the situational report of every OSEP office in Tel Aviv and Jerusalem

Msc.ing Feuzeu Yomsi
Raymond Jacquino

X1.C.ANALYSE OF OBAMA STIMULUS EDUCATION PLAN EFFECT IN 4 YEARS In ISRAEL

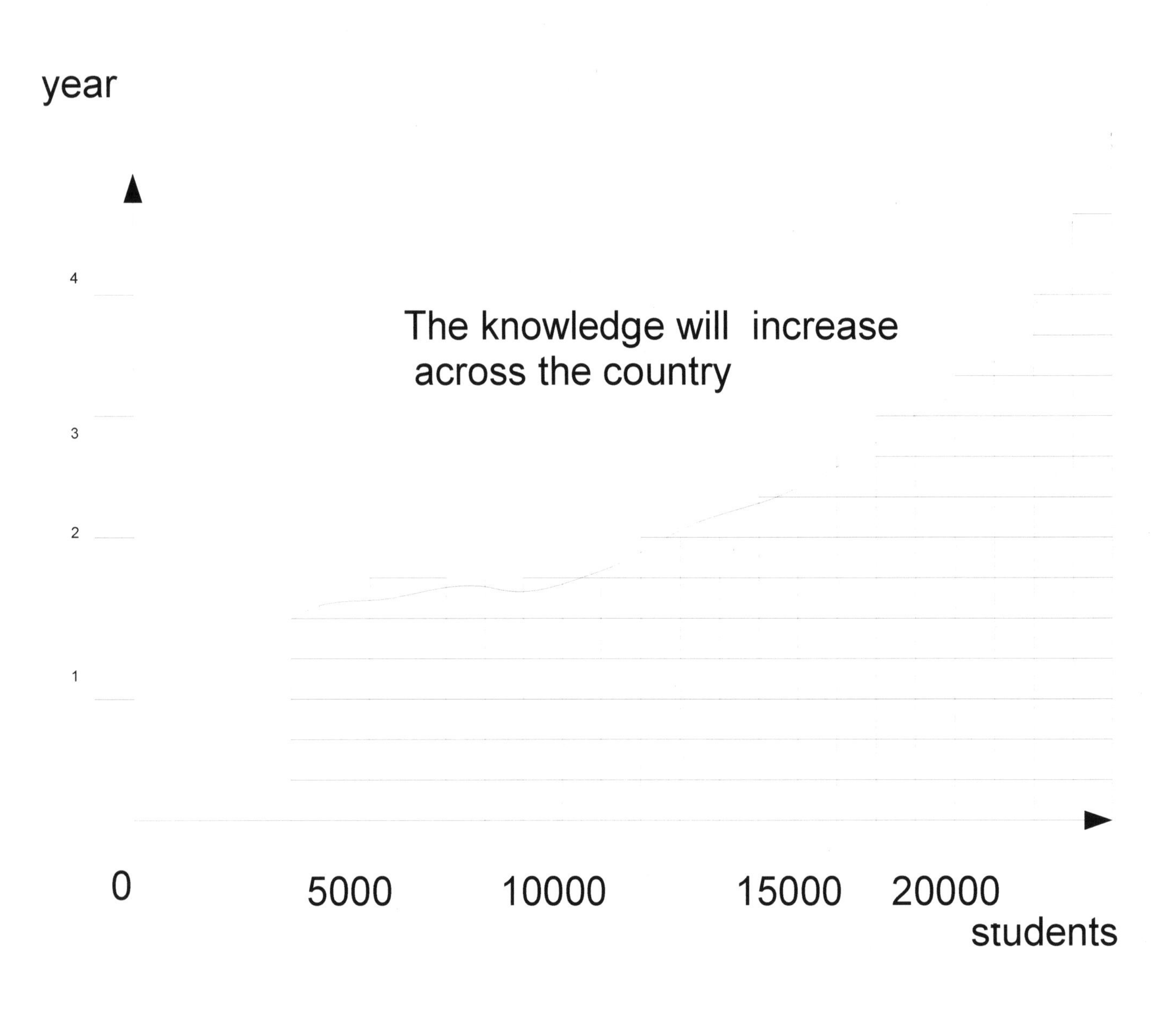

Msc.ing Feuzeu Yomsi
Raymond Jacquino

XI. TEST AND ANALYSE OF THE OSEP IN GERMANY

A. Entry Data

a) number of City =2 :BERLIN and HAMBURG

1. GERMANY
   city : BERLIN
2. GERMANY
   City :HAMBURG

b) number of OBAMA STIMULUS EDUCATION office in GERMANY=5

OSEP/GERMANY/BERLIN/office1=id1=100a
OSEP/GERMANY /BERLIN/office2=id2=101a
OSEP/GERMANY/HAMBURG/office1=id3=200b
OSEP/GERMANY/HAMBURG/office2=id4=201b
OSEP/GERMANY/HAMBURG/office3=id5=202b

c)number of students who receive a consultation at OBAMA stimulus education plan office in Berlin and in Hamburg = 10

d)students receive appropriate consultation in :

1.mathematic in Berlin
2.Information technology in Hamburg

B. The central administration in GERMANY need the situational report of the OSEP across the country in this particular case in BERLIN and HAMBURG
Eventually needed to know the average of satisfy students or not satisfy by the OSEP.

Our goal is to minimize the academy difficulties face by the students, maximize the success , increase the level of knowledge .

C. Graphical Interpretation

XI.TEST AND ANALYSE OF THE OSEP EFFECT In GERMANY

OSEP
GERMANY

OSEP
GERMANY

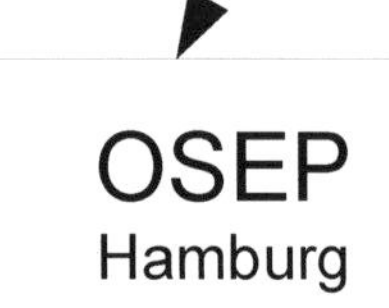

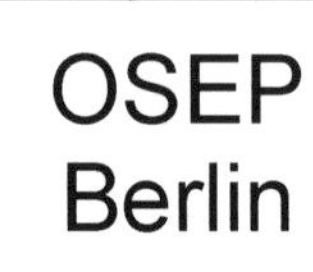

The OSEP central control in Germany Need the situational report of every OSEP office in Hamburg and Berlin

Msc.ing Feuzeu Yomsi
Raymond Jacquino

X1.C.ANALYSE OF OBAMA STIMULUS EDUCATION PLAN EFFECT IN 4 YEARS In GERMANY

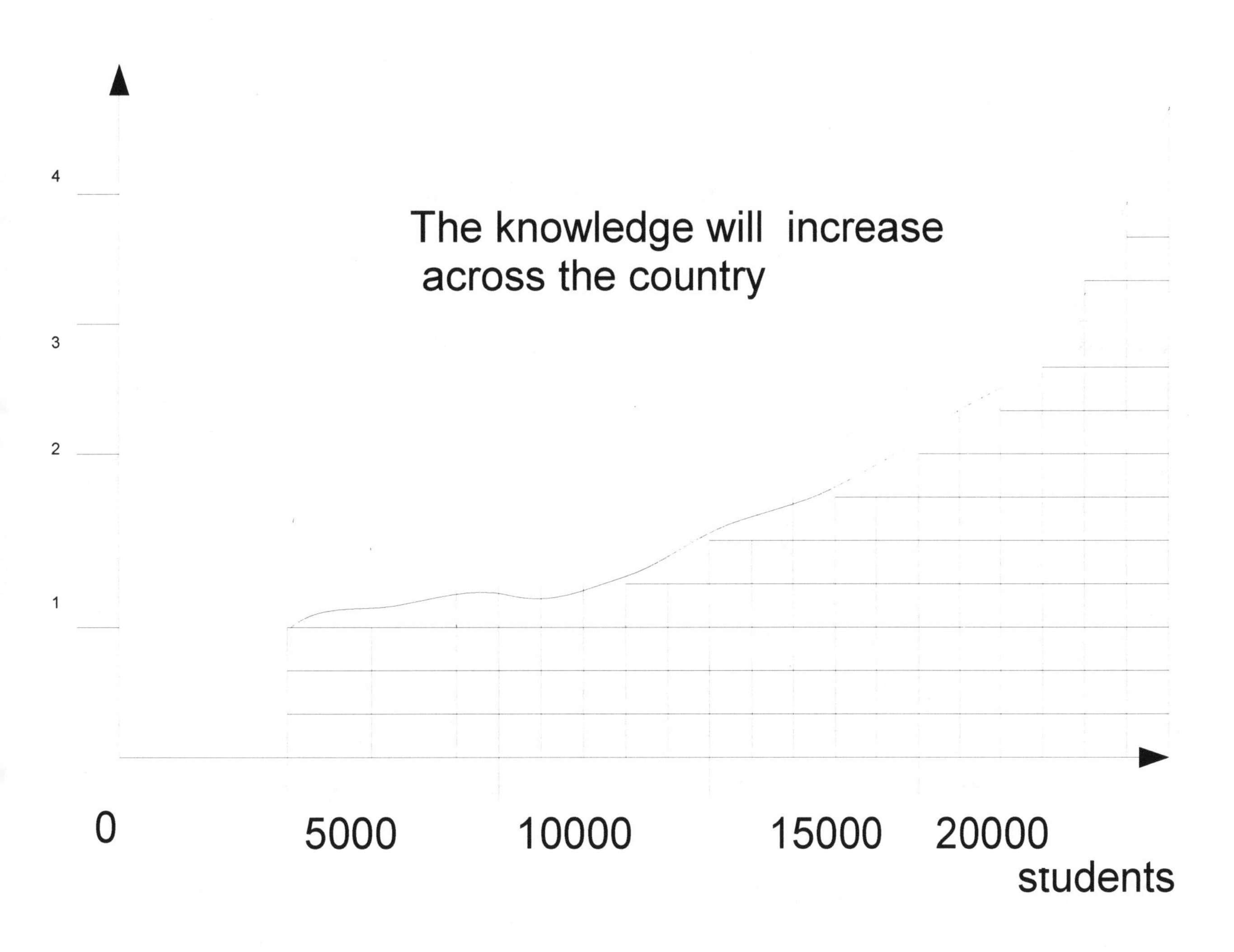

Msc.ing Feuzeu Yomsi
Raymond Jacquino

APPRECIATION

Msc.-ing Feuzeu Yomsi Raymond Jacquino Master in Engineering in information processing control . Education is power and I have the honor to design this book to help our President BARACK OBAMA. The plan if adopted will be named: OBAMA STIMULUS EDUCATION PLAN. The plan will challenge the education system, will help thousands of Kenyan students to became successful in all academic areas by putting in place a Network of tutoring to assist students, the plan will open windows of opportunity to children, restore prosperity and make Kenya progress.

The main recommendation of this plan is that the students be strongly encouraged to indicate the efforts and attempts made to resolve a particular problem. Actually, the system is not meant in anyway to encourage laziness on the part of the students.

It is recommended here as pilot plan with the hope that this would be tested in Kenya, USA, Cameroon, France, Russia, Indonesia,India,China, Japon, England, Brasil ,Argentina, South Africa,Israel Germany and why not elsewhere in the world where need be. Given the cultural differences between countries as well as the differences in educational systems and infrastructural setting, this is not expected to be a fit-all plan. In effect, there is need to take into the account the cultural realities of each country both in the testing and implementation of this plan, and adjusting this to meet the special needs of each setting

Author:
Msc.-ing Feuzeu Yomsi
Raymond Jacquin
The president of the Cameroonian Association in Greater Philadelphia

IN GOD WE TRUST